The Impact of Policy Reform and Institutional Transformation on Agricultural Performance

DEVELOPMENT ECONOMICS AND POLICY

Series edited by Franz Heidhues and Joachim von Braun

Vol. 47

PETER LANG

Frankfurt am Main · Berlin · Bern · Bruxelles · New York · Oxford · Wien

The Impact of Policy Reform and Institutional Transformation on Agricultural Performance

An Economic Study of Ethiopian Agriculture

Beyene Tadesse Ferenji

PETER LANG
Europäischer Verlag der Wissenschaften

Bibliographic Information published by Die Deutsche Bibliothek
Die Deutsche Bibliothek lists this publication in the Deutsche Nationalbibliografie; detailed bibliographic data is available in the internet at <http://dnb.ddb.de>.

Zugl.: Hohenheim, Univ., Diss., 2004

D 100
ISSN 0948-1338
ISBN 3-631-53577-5
US-ISBN 0-8204-7691-9

www.peterlang.de

To my children
Fraol and Lelisse

Preface

From variety of economic development activities we have learned that a conducive economic policy and institutional framework are of crucial importance for economic development. Many developing countries have launched remarkable policy reforms leading their economies toward market oriented economic systems primarily aiming at promoting sustainable economic growth and reducing poverty. The pursuit of such reform policies, however, is successful only if infrastructure and institutions are in place and developing.

It is evident that agriculture dominates the economic lives of most people, particularly the poor, in less developed countries. Hence, it is important to give due emphasis to agriculture to promote increased food production, reduce poverty and facilitate the transfer of resources necessary for the expansion of non-agricultural sectors; productivity growth in agriculture is the key factor in promoting the process of structural transformation. This volume of the series *Development Economics and Policy* discusses the impact of Ethiopia's economic reform process on the overall performance of the Ethiopian economy with a special focus on agriculture and food production. Using a Stochastic Production Frontier combined with an Error Correction Model the study analyzes the determining factors of technical progress, level of technical efficiency in factor use and their interaction with total factor productivity. Important for policy formulation is the study's finding that in production sectors with rapid technical progress (such as in maize production) farmers' knowledge and the institutional support systems needed for applying the innovations turned out to constrain the increase in total factor productivity. In traditional sectors (such as tef production) marketing systems was less constraining than technical innovations. Thus, formulating effective productivity supporting policies requires a differentiated approach to allow for targeting at key 60constraints.

The analysis also indicates that farm product prices have become more volatile after the reform; higher price fluctuations increase uncertainty and result in considerable negative impacts on farmers' incentive to use modern inputs. Thus they hindered growth in domestic food supply.

The study concludes that policy reform in the course of fundamental structural adjustment, such as in Ethiopia, has a chance of merely limited success in transforming agriculture if the strengthening of institutions and markets, particularly research and extension services, is not proceeding in parallel. Both institutional reforms and policies to expand markets and agricultural support systems may need to be differentiated by sector and crop depending on where the key constraints are found. Advanced econometric models can make important contributions to well targeted policy design.

Prof. Dr. Franz Heidhues
Center for Tropical and
Sub tropical Agriculture,
University of Hohenheim

Prof. Dr. Joachim von Braun
International Food Policy
Research Institute (IFPRI),
Washington D.C.

Acknowledgement

A number of people and organization contributed to the successful completion of this work, and I thank them all. First and foremost, I would like to express my heartfelt gratitude to my supervisor Prof. Dr. Franz Heidhues for his guidance, encouragement and his readiness to give me professional advices. He has also given me several opportunities that assisted me to widen my understandings of the development theories and policies. I am also most grateful to my co-advisor Dr. Mulat Demeke who made a valuable contribution to the first outline, and later provided penetrating comments on the drafts that helped me address most pressing issues of the time in Ethiopia. My special thanks are due to Dr. Assefa Admassie and Dr. Woldy Amaha for their motivation and preparation of the ground to embark on such a formidable writing task. I have special appreciation to Dr. Andre Croppenstedt and Dr. Maja Micevska for their careful comments on the issues and methodologies of the research. I am duly grateful to Mr. Negussie Tefera, Bekele Hunde and Tadele Ferede for their strong support in the preliminary processes of the research data. The entire text was read by Mrs. Anna Canning who did an excellent job of editing the final draft. I am also indebted to Mr. Clemens Breisinger for translating the summary part into Deutsch. Without the genuine support of all these scholars the work would have not been successfully completed.

I wish to thank the staff members and students at the Department of Eonomic Development Theory and Policy (Institute 490a), University of Hohenheim, for their unreserved supports. My profound appreciation goes to Dr. Gertrud Buchenrieder, Mrs. A. Shempp and G. Contag for their assistance in all administrative processes and moral supports during my study. Special thanks go to my colleagues Mrs. Sabine Daude, Mrs. Judith Möllers, Mr. Rainer Schwarzmeier, Ms. Dinh Ngoc Lan and Ms. Pamela Ayiera, and my friends Dr. Eshetu Beshada, Mr. Teklu Erkossa, Mr. Dereje Woltedji, Mr. Hailu Adugna and Ms. Martha Oumer for their continues encouragement and lovely time we had together.

I duly acknowledge the German Academic Exchange Services (DAAD) for offering me the scholarship. The research was funded by DAAD, Universitätsbund Hohenheim and Daimler Chrysler. I also wish to thank the Department of Economics of Addis Ababa University for the opportunity it provided me to access and use its rural and urban databases.

I express my sincere thanks to my wife Yehualashet Gobena and my children Fraol and Lelisse for their patience and tolerance of the hardships they confronted with and the affection they missed because of my absence. Thanks to the Almighty Lord for the endurance He has given us.

Beyene Tadesse

Table of Contents

List of Tables

List of Figures

Acronyms and Abbreviations

ADF	Augmented Dickey-Fuller
ADLI	Agricultural Development Led-Industrialization
ANOVA	Analysis of Variance
CD	Cobb-Douglas
CSA	Central Statistical Authority, Ethiopia
DC	Developed Countries
EPRDF	Ethiopian People's Revolutionary Democratic Front
FAO	Food and Agricultural Organization of the United nations
GARCH	Generalized Autoregressive Conditional Heteroscedasity
GDP	Gross Domestic Product
ICRAF	International Center for Research in Agroforestry
IFPRI	International Food Policy Research Institute
IMF	International Monetary Fund
LDC	Less Developed Countries
LR	Generalized log Likelihood Ratio
MEDaC	Ministry of Economic Development and Cooperation
ML	Maximum Likelihood
MOA	Ministry of Agriculture, Ethiopia
OLS	Ordinary Least Square
PADETS	Participatory Demonstration and Training Extension System
PP	Pillips-Perron
PPF	Production Possibility Frontier
RTS	Returns to Scale
SAP	Structural Adjustment Program
SD	Standard of Deviation
TCH	Technological (Technical) Change
TE	Technical Efficiency
TEC	Technical Efficiency Change
TFP	Total Factor Productivity
VECM	Vector Error Correction Model

1 Introduction

1.1 Background

The Ethiopian economy, like that of many sub-Saharan African countries and other developing countries, is based predominantly on agriculture. The agricultural sector accounts for about 47% of the gross domestic product (GDP) and about 85% of exports (CSA, 2000). It supports a substantial proportion of the labor force, an estimated 85% of the total population of about 67 million. The growth performance of the agricultural sector could therefore to a large extent determine the process, pace and prospects of development in the non-agricultural sectors and the economy as a whole. However, the other sectors, which support only 15% of the total labor force, produce 53% of the country's GDP.

About 70% of agricultural GDP comes from crop production, while the remaining 30% is from livestock and other agricultural activities (CSA, 2003). Subsistence and mixed farming systems characterize the sector. Consequently, the sector is dominated by small-scale farmers cultivating about 96% of the total area under crop, and producing more than 90% of total agricultural output and 97% of food crops. Land, traditional tools, draft animals and family labor are still the most important factors of production. Oxen power is critical in determining how a farm operates, and a shortage of oxen power significantly reduces both the area of land cultivated and the yield, and hence total farm income (Assefa, 1995). Compared to global agricultural practice, the use of chemical fertilizers and improved seeds at national level is negligible even by African standard. Crop production is almost entirely rain-fed, and irrigation is rare.

Ethiopia has at least moderate agricultural potential, but with enormous degrees of variation across ecological zones. Potential ranges from very high in broad areas of the central and south-western zones, to very low in the arid zones of the north and east. According to the results of the MEDaC (1999) survey, the country has about 112 million hectares of land, of which 65% is known to have agricultural potential, although to date only 22% has been used. The land is endowed with different soils and climates that allow cultivation of a variety of annual and perennial crops, and rearing of a variety of classes of livestock. There are also numerous water bodies suitable for irrigation purposes. The irrigation potential of the country is estimated at 3.5 million hectares, but very little (< 5 %) of this has been exploited.

Despite its paramount importance in the economy and its potential for growth, however, Ethiopia's agricultural sector has performed poorly in terms of both production and productivity. The problem was particularly serious throughout the entire Socialist Derg era (1974-1991). Productivity was a mere 1.0-1.5 tons per hectare at national level (CSA, 2000). Agricultural food production and the supply of raw materials to the industrial sector thus remained well below the national requirement. Hence, the nation experienced poor overall economic performance, with the result that about 40% of the population lived below the poverty line. Consequently, critical food shortages and recurrent famine led to a need for massive food aid and imports (FAO/UNDP, 1984; FAO, 2002). The macroeconomic indicators for the country during the Socialist regime reveal a disquieting agricultural food production rate of 1.75% at a time when the population was growing at the rate of 2.9%, implying a food deficit. In the same period, per capita GDP fell by 15%; domestic savings declined by 48% and fixed capital formation by 30%. Both internal and external macroeconomic imbalances provided a further indication of the frustrating state of Ethiopia's economic performance (Hamza et al., 1995). The budget deficit during 1989/90 amounted to 17% of GDP, while the annual inflation rate hit 22%. More importantly, the poor performance of agriculture was most clearly evidenced by low standards of living in the rural areas, where the largest concentration of the population (47.5%) lived below the poverty line, and malnutrition and infant mortality were high (Abebe Shimeles, 2004).

Several explanations may be given as to why the agricultural sector performed so poorly. The causes of the problem were manifold: inappropriate policy formulation, weak technological progress, lack of infrastructure and absence of proper institutional frameworks are frequently acknowledged as contributing factors. Population pressure and repeated occurrence of unfavorable weather conditions may also have had immeasurable effects (Dejene, 1997; Tegegn, 1995). Although no quantified evidence is available on the relative importance of the causes of poor performance in the agricultural sector, many of the drastic effects were attributed largely to the inadequacy of the economic policies and strategies implemented by the Socialist regime. Pricing and marketing policies were aimed at stabilization, using strategies such as fixing producer prices and restricting inter-regional movements of grain. The policies of that time were geared towards protecting urban consumers' welfare by imposing direct controls to keep prices low, and indirect policy measures too had an adverse impact on rural producers. Private sector activities were deliberately suppressed; instead, producer cooperatives and state farms were favored. Consequently, public policy adversely affected the performance of the agricultural sector, resulting in a fall in farmers' real incomes due to distorted rural-urban and international terms of trade (TGE, 1991). The development policy of the Socialist regime was thus effectively biased against agriculture and neglected the rural sector as a source of growth (Martin and Mitra, 2001).

Since agriculture is the mainstay of the economy, however, it should attract major political and policy interest. The dismal performance of the Ethiopian economy dictated to the present government (which came to power in 1991) the necessity of undertaking sweeping policy reforms to replace the former centrally planned economy with a market-oriented economic system. A Structural Adjustment Program (SAP), initiated by the World Bank and the International Monetary Fund (IMF), has been undertaken stepwise since 1992. It encompasses all-round components, such as reducing/switching expenditure and institutional reforms. Market liberalization (allowing strong competitive pressure from internal and international markets), devaluation, privatization of productive resources and elimination of subsidies (said to have created hard budget constraints) are among the major reforms accomplished. Moreover, investment in rural productive infrastructure and institutions has been a top priority on the agenda (EPRDF, 1996). As is repeatedly advocated, the ultimate objective of economic and institutional reform is to encourage competition among economic agents, improve efficiency of production and thereby foster economic growth.

However, since the March 1975 land reform proclamation of the Socialist Derg regime, land ownership has remained under state control, and is subject to redistribution. Based on the three main types of informal and less secure land contracts (rented, share-cropped and borrowed), Gavian and Ehui (1999), have argued that there is no empirical basis to support the claim that land tenure in Ethiopia is a constraint to agricultural productivity. Nevertheless, the need to privatize land is still a hotly debated policy issue.

In Ethiopia, the growth of the agricultural sector is particularly important, as it constitutes the bulk of the economy in terms of both human and material resources. Agriculture is therefore required to show dynamism and growth if it is to alleviate poverty and to promote economic growth as a whole. Primarily, it has to satisfy the food demand of the rapidly growing population and release an adequate surplus to the other sectors of the economy. In order to achieve food self-sufficiency and then accomplish structural transformation of the economy, the development strategies of the nation therefore need to give due emphasis to agriculture. With this in mind, the government formulated a rural-based economic development strategy termed "Agricultural Development-Led-Industrialization" (ADLI) in 1992/93. As the strategy's title implies, agriculture was supposed to play a dynamic role in satisfying the food requirements of the nation and paving the way for industrialization. The cornerstone of the strategy was primarily to improve the productivity of smallholders in order to increase food supply and secure the country's food requirements, and then to boost industrial growth utilizing agricultural raw materials. The strategy correlates with the notion that an effective economic development strategy depends critically on promoting productivity and output growth in the agricultural sector, particularly among small-scale producers (Johansen, et al, 1961; Hayami and Ruttan, 1985). Thus, improvements in resource use efficiency and technological progress are indispensable. To this effect, a new agricultural extension strategy entitled "Participatory Demonstration and Training Extension System" (PADETES) was initiated in late 1994 as a key instrument for promoting agricultural growth (Belay, 1997). Massive diffusion of agricultural technologies (mainly improved seed, fertilizer, pesticides and herbicides) was initiated from that time in almost all parts of the country.

Policy reforms and institutional changes were expected to boost agricultural output. At the early stages of growth, many of the effects could be seen at micro or farm household level. Change in agricultural output is presumed to have a subsequent spill-over effect on the other sectors of the economy through the forward and backward linkages at the macro-level as well. It is therefore crucially important to have an in-depth understanding of the pace of agricultural output growth and factors contributing to it. Moreover, the stability of output growth and market prices is another aspect that must be evaluated when reviewing the performance of market-based development policies. Because fulfilling reliable food surplus is a prerequisite for economic development, the study focuses on major sources of food in Ethiopia.

1.2 The Issues

The radical changes accompanying economic transition in the early 1990s in Ethiopia were supposed to initiate sustainable economic growth to improve the living standard of society as a whole. Reforms such as eliminating agricultural support, market liberalization and privatization of resource ownership are frequently advocated for bringing about more efficient resource use, particularly if complemented with infrastructure improvements and compatible institutional changes. In general, the overall potential of the agricultural sector as an engine of growth and development hinges upon the performance of markets and stability of market prices to guide efficient resource use, assuming that good governance and a stable political environment prevail in the economy.

There remains a significant gap between expected and actual levels of efficiency and growth, however, and this is proving problematic for the economic agents facing international competition on the liberalized markets. Institutional shortcomings or failures such as market uncertainties and incompleteness, lack of credit institutions, and information asymmetries negatively influence the efficiency of economic agents. The gap between the perfectly competitive cost-free market of neoclassical theory and real markets in the country should reduce our expectation that firms perform on their efficiency frontiers, as well as our expectations of their effectiveness at sectoral level and expectations of overall economic growth.

There are two conventional, non-exclusive ways of increasing agricultural production. The first is intensification or expansion of resource use, particularly by bringing more land under cultivation and making use of the abundant rural labor force; the second option is to increase the productivity of farm resources (through technological and efficiency improvements). Although the first option is likely to be feasible only in the short term from the sustainability point of view (because land is limited), studies have shown that the latter option is encouraging, and gains are enormous (Schultz, 1964; Timmer, 1988). Initiating and managing such a strategy, however, is perhaps the most difficult task facing developing countries in general, and especially those countries which, like Ethiopia, have experienced persistent stagnation or even decline in their economies. This is because of the slow nature of progress in technological, institutional and infrastructure development, which in turn significantly influences market efficiency and resource use.

In addition, studies on economic growth argue that agro-climatic conditions and the resource base are crucial in determining agricultural potential. As was explained in brief above, Ethiopia has quite diverse agro-ecological zones with substantially varying agricultural potential. These variations significantly influence growth in agriculture through their effects on choice of and access to technologies and infrastructure. Also, soil degradation has increased considerably, resulting in a terrible loss of nutrients (and, by implication, depreciation of natural capital). Agricultural production is still heavily dependent on rainfall and is thus very vulnerable to weather changes and other risks and uncertainties. It is also often argued that product markets are underdeveloped, while factor markets lag even further behind. Altogether, this situation can hamper the potential for efficiency improvements and technological progress in the country. Consequently, agriculture may be less responsive to policy changes in Ethiopia, and it may even be inflexible when confronted with an unstable political environment. According to the recent study on governance by representatives of the United nations' Economic Commission for Africa (Okey, 2004), the government of Ethiopia has been ranked 3rd at the bottom compared to even other African countries scoring too far from the sample average. Bad governance in general has been found to have significant negative effects on efficiency and growth prospects (Schlüter, 2001; Boger, 2001).

Many studies relating to agriculture in Ethiopia have focused on adoption of improved technologies, and only a few (e.g. Assefa, 1995; Gavian, et al., 2002) on production efficiency and productivity. Both types of study are based on cross-sectional data (perhaps due to a lack of time series data). The results obtained in these studies were only able to address the existing situation regarding a particular component of growth (e.g., use of modern inputs, or efficiency) in a particular cropping season. The available information on the performance of the agricultural sector in the post-reform period is therefore only partial. There is a lack of information on the overall performance of agriculture over time, which is important for designing appropriate policy measures.

In order to evaluate the impact of the structural adjustment program on agricultural growth, many scholars focus more on analysis of agricultural supply response (Schiff 1997; Junankar, 1989; Binswanger, 1989; Sadoulet et al., 1994) while others concentrate on its impact on income distribution. But in a poor country like Ethiopia, where the market is not so highly developed and where the rural people are poor at large, we believe that a more comprehensive approach for evaluating the impact of policy reform and institutional change is to examine its impact on the overall components of agricultural growth. In other words, simultaneously evaluating trends in resource use intensity, technological change and improvements in efficiency of production over a certain period.

This would answer some fundamental questions: are market prices guiding farmers to use resources more intensively and efficiently, and are there incentives to use more modern inputs on a sustainable basis? What development is taking place in terms of technology and efficiency of production, and where is there still room for improvement? Which components (sources) of output growth have been more affected by the reform in terms of contributing to or limiting output growth? Considerable time has passed since the reform was introduced, and so, under normal circumstances, tangible differences should be discernible in the performance of the economy in general and that of the agricultural sector in particular.

1.3 Hypotheses

Acknowledging the notion of institutional economics that underlines the importance of institutions and infrastructure development for successful economic growth, the objective is to assess the performance of the agricultural sector in the post-reform period. It is now more than ten years since the beginning of the transition. Economic actors responding to the reforms are thus expected to have progressed in terms of active and effective economic activities. Given that technological advancement in Ethiopia is far behind that of the developed countries, and assuming that agriculture is in continuous processes of transformation, the main hypotheses of the research are postulated as follows:

1. The shift from a planned to a market-based economic system should lead to a more efficient economy by providing producers with incentives to realize input-output combinations at the production frontier rather than inside the production possibility set. Thus, the mean technical efficiency of small farmers producing food crops is expected to increase over time assuming less distorted (better functioning) institutions. Improvements in mean technical efficiency prevail particularly if no large differences in production technologies emerge. On the other hand, however, technical efficiency could remain constant or even decline, despite positive market incentives, if only the best farmers implement rapid technological change and a large number of farmers are left behind. In addition, technical efficiency may be higher for farmers located in areas that have better access to markets and transport networks than those in inaccessible areas. This difference is expected to be minimized after the reform.

2. Heavy involvement of government extension services in the diffusion and transfer of improved agricultural production technologies to rural farmers is expected to bring about technological progress that may be manifested in a shift in farmers' production frontiers from their initial level. Improvements in production technology should also reduce the

productivity gap between areas with different agro-ecological potential, whereas at the outset productivity was entirely dependent on natural factors. Success depends on the effectiveness of the technological packages and the extent to which a significant number of farm households are willing and able to adopt the improved technologies in a sustainable manner.

3. The cumulative effect of changes in technical efficiency and technological progress is expected to improve total factor productivity in food crop production.

4. Because the product market is free from government intervention, its stability depends to a large extent on the condition of the domestic economy and partly on the international market. After the reform, market prices could become more volatile. This would lead to a greater price risk for farmers and so reduce the incentive to use modern inputs consistently, resulting in unreliable food grain supply.

1.4 Objectives of the Study

The general objective of the research is to examine the impact of policy reforms and institutional changes on the performance of agriculture in Ethiopia. More specifically, this study attempts to address the following interrelated objectives:

1. to describe changes in farmers' resource endowment (physical capital formation), changes in factor market, use of modern inputs and the production pattern of major crops;
2. to measure the extent to which technological progress, technical efficiency and intensity of input use have changed since the reform, and to estimate their relative contribution to growth in agricultural production;
3. to analyze trends in total factor productivity vis-à-vis growth in agricultural production, and explain how the reforms have affected this by influencing the incentive structure for promoting efficiency and technological change;
4. to assess the price stability of major agricultural products and the impact this has on modern input use and food grain supply, which in turn has important implications for sustainable use of improved technologies and growth in agriculture; and
5. to identify major factors determining more rapid improvement in technical efficiency and technological progress.

1.5 Significance of the Study

Policy reforms and institutional changes are the result of conscious economic policy choices. Considerable funds from both domestic and external sources have been committed for their design and implementation in the economy in general and with especial emphasis on agriculture in particular. Understanding the pace of agricultural growth and information about growth in the productivity of the sector is thus imperative. It is instructive for policy making with regard to food security, and for forecasting the road ahead to structural change that is so central to the development process of agriculture-based industrialization.

The study provides insights into the trends and the extent to which farmers' productivity and output have changed, and explores how stability in output prices affect farmers' incentives to use modern inputs. Quantifying the magnitude of the partial contribution of each distinct source of growth highlights the effect of policy and institutional reforms on farmers' production. Information on the dynamics of technological progress and efficiency is essential for an economy that is striving to facilitate the process of agricultural transformation. Moreover, the role of attractive and reliable (stable) market prices in encouraging farmers to use modern inputs on a sustainable basis and inducing increased output supply merits policy attention. Therefore, empirical tests of the hypotheses and detailed analysis of the position of farmers' production in a market-based economic system are important and instructive for policy-makers and planners to undertake appropriate intervention to speed up growth in the sector.

1.6 Structure of the Thesis

The thesis is organized into seven chapters. In *Chapter Two*, a synopsis of development theories and experiences is presented. Starting with a brief discussion of the major arguments against agriculture, it goes on to provide an insight into the role that agriculture can play in economic development, with supporting evidence from the success of current developed and transition economies. The problems and failures of the sub-Saharan African countries in their efforts to enhance agricultural development are also discussed in this chapter. In addition, the chapter also elucidates the essential determinants of agricultural growth, such as the role of the resource base, technical progress, effective demand, and policy and institutional factors. *Chapter Three* is taken up exclusively with an overview of the economic background of Ethiopia, briefly summarizing economic development policies in general and policies and strategies concerning the agricultural sector in particular, both before and after the reform. It also presents a description of changes in trends in the main economic indicators, with a special focus on agriculture and food production.

The remaining sections of the thesis are concerned with empirical analysis relating to the objectives and hypotheses discussed in the previous sections. *Chapter Four* provides a micro-level analysis of the magnitude of changes in agricultural productivity and output growth after the reform, using panel data from farm households in different agro-ecological zones. It investigates the significance of changes in technical efficiency, technological progress and total factor productivity and their relative contribution to output growth, using a growth decomposition approach. Based on the results of this chapter, *Chapter Five* goes on to examine constraints to improving technical efficiency in a selected production line. *Chapter Six* explores the degree of instability in product prices and provides an econometric analysis of its subsequent impact on the utilization of modern inputs and on the supply of major food grains at the macro level. Finally, *Chapter Seven* presents the summary, conclusions and policy recommendations of the study.

2 Synopsis of Development Theories and Experiences: the Significance of Agricultural Transformation and its Roles in the Dynamics of Economic Growth

Lewis (1954) first introduced the idea of structural change[1], nowadays referred to as structural transformation. Schultz (1964), defines transformation as change that occurs in response to new economic opportunities, and which originates from improvements in the state of the productive arts. Structural transformation includes virtually all economic functions, but primarily involves the transformation of production. In other words, it entails a decline in the relative importance of agriculture and an increase in that of the industrial and service sectors of an economy, accompanied by a shift in resources, particularly that of labor, in the same direction. Secondly, it involves changes in the composition of consumer demand. Demand for food declines relative to the demand for non-food consumption. At the aggregate level, the relative importance of investment, government expenditure and international trade (exports of manufactured goods and services) increase. Furthermore, changes occur with regard to socioeconomic factors such as urbanization and the demographic structure (Staatz, 1998; Todaro, 1989; Kuznets, 1973).

The production component of structural transformation is the driving force behind other changes. This has been attracting the attention of many scholars. To put it more specifically, the economy becomes relatively less agriculturally oriented, although agriculture and, more broadly, the food system continue to grow in absolute terms and generate important growth linkages to the rest of the economy. Structural transformation thus involves *a net resource transfer from agriculture to other sectors of the economy, over the long term.* Therefore, the focus is on the mechanisms by which underdeveloped economies transform their domestic economic structure from predominantly rural and traditional subsistence agriculture into a more modern, more urbanized, and more industrially diverse manufacturing and service economy. This structural transformation has involved expansion of non-farm employment, increased integration of agriculture with the rest of the economy, and expansion of off-farm elements of the food system. Two well-known representative examples are Lewis's theories and Hollis Chenery's empirical analysis (Chenery, 1960).

[1] In Lewis's model, the underdeveloped economy consists of two sectors: 1) a traditional, overpopulated rural subsistence agricultural sector, characterized by zero marginal labor productivity that can be withdrawn and transferred to the modern sector without any loss of output in the agricultural sector, and 2) a high-productivity modern industrial sector into which labor from the subsistence sector is gradually transferred. The structural transformation of the economy takes place with the balance of economic activity shifting from traditional rural agriculture to the modern urban industry (Todaro, 1989).

Economic development is the outcome of the structural transformation process (Todaro, 2003). If so, what is (are) the most appropriate development approach(es) in order to turn an underdeveloped nation into a developed one? Is there a dynamic relationship between agriculture and industry that reinforces an optimal growth strategy? In Ethiopia, there is ongoing debate among development practitioners, professionals and policy-makers as to whether ADLI is an appropriate development strategy that will lead to industrialization and generate economic growth. This chapter, which draws on the principal theories of development and on evidence of success and failure experienced by a wide range of countries, aims to contribute to this debate. The role of agriculture in economic dynamism and mechanisms for accelerating its growth are briefly summarized.

2.1 Agriculture in Historical Perspectives of Development Process

Up until the late 19th century, the role of agriculture in the development process was not given much attention, and it remains contentious even now among some scholars of developing countries. The mainstream paradigm in the 1950s generally suggested that agriculture could and should be squeezed for surplus on behalf of the more dynamic sectors of the economy. This strategy, however, could be useful only if agriculture was already growing rapidly (as in Western Europe and Japan), or if it started with a large surplus relative to the subsistence needs of the rural population (as in the early USSR). Timmer (1998) therefore argues that if traditional technology is predominant in the agricultural sector and living standards are near subsistence levels, the "squeeze agriculture" paradigm creates economic stagnation, not growth. Rather, he argues, focused efforts are needed to induce agricultural transformation if an industrial revolution is to take place successfully. The three most common arguments against agriculture are discussed below.

2.1.1 The Dualistic Model Argument

Classical and the neo-classical development approaches have given most emphasis to industry and often neglect the role of the agricultural sector. For instance, the Soviet Union's practice in the 1920s was equated with industrialization with modernization. Capital and labor were believed to be more productive in industry than in agriculture. This is because industry was seen as having major economies of scale and external economies. Then it was hypothesized that increases in the capital stock were the source of growth, and that this resource should be directly allocated to production of capital goods rather than consumer goods, including agricultural production. This is because agriculture was viewed as backward and static and subject to diminishing returns.

They used the concept of dualism[2] to explain the vitality of the industrial sector and the secondary role of the agricultural sector in development processes. The main concerns of the classical and the neo-classical approaches in the development of the dual economy were the technology of the agricultural sector and the conditions governing the supply of labor. It is assumed that there is some level of the agricultural labor force beyond which further increments in this labor force are redundant. This labor can be available to the industrial sector in unlimited supply (infinitely elastic) at a fixed real wage rate set above average farm income. In other words, the marginal productivity of labor in the agricultural sector is zero or positive, but less than the real wage. This implies that labor could be supplied to the industrial sector with little or no loss in terms of agricultural output.

Lewis' (1954) seminal article *"Economic Development with Unlimited Supply of Labor"* strongly influenced development economists throughout the 1950s and 1960s. Lewis argued that no relationship exists between the activities of backward agriculture and the modern industrial sectors. According to him, a backward agricultural economy eventually develops into an advanced industrial economy; the boundary between the two is marked by the disappearance of disguised unemployment in the former sector. His model therefore focused on how to facilitate the transfer of labor from the subsistence sector to the capitalist industrial sector. In Lewis' model, expansion in the capitalist sector continues until earnings in the two sectors are equated, at which point a dual sector is no longer relevant; growth proceeds as in one sector. The classical approach further assumes that financial savings are equal to total profits in the industrial sector and that all is done by people who receive profits or rents in the industrial sector. The availability of unlimited labor at constant real wages raises the capitalist surplus continuously and annual investment will rise as a proportion of the national income (Lewis, 1954; Fei and Ranis, 1964). That is, in each case the explanation of growth is based on the relationship between savings and industrial profits.

The theories led many policy-makers and planners in underdeveloped countries to focus on industrial development. Only a few theorists incorporated the production side of agriculture into their models. The latter primarily proposed mechanization and land consolidation as a means of increasing

[2] Dualism represents the coexistence of two extreme conditions with increasing divergence. The popular example in development economic theory emanates from the Lewis' model of the coexistence of modern capitalist methods of production in the urban environment and traditional, subsistence methods of production in rural sectors. According to the dependency notion, dualism represents the coexistence of powerful and wealthy industrialized nations with weak, impoverished peasant societies in the international economy.

agricultural productivity, although these require drastic structural changes in agriculture and presuppose a massive absorption of surplus agricultural labor by prior industrialization. Jorgenson (1961), on the other hand, argued that disguised unemployment is neither a necessary nor a sufficient condition in itself to generate a sustainable rise in the share of savings. A sustained increase in savings depends on a positive and growing agricultural surplus, and not on the presence or absence of disguised unemployment. Assuming diminishing returns, the classical and neoclassical theories paid inadequate attention to the need for technical change in agriculture and to the biological and location-specific nature of the agricultural development process (Nicholls, 1964).

2.1.2 The Terms of Trade Argument

Export earning is one of the most important components decisive for economic growth, which in turn is greatly determined by terms of trade. The relative neglect of agriculture in the 1950s was reinforced by Prebisch (1950), who argued that, over time, 'terms of trade' would become detrimental to countries that export primary products and import manufactured goods. They concluded that the scope for growth through agricultural and other primary exports was very limited. The view that primary commodity prices, particularly those of agricultural commodities, would inevitably show a downwards tendency relative to the prices of manufactured goods led Prebisch (1950) to espouse an import substitution strategy for Latin America and elsewhere. This requires developing countries to shift out of agriculture and into industry as quickly as possible. Moreover, almost all rich countries protect their agricultural sector and their key food production from international competition (World Bank, 1986) by directly restricting imports and/or by subsidizing their farmers. This absolutely limits the scope of developing countries to benefit from exports of agricultural products.

2.1.3 The Relative Importance Argument

The faster agriculture grows, the faster its relative size declines (Mellor, 1998; 1995). The decline in importance of agriculture in a country's labor force and total output in the development process is slightly puzzling when viewed vis-à-vis the apparent prerequisite of rapid agricultural growth. This apparent paradox gave rise to the commonly held misperception that agriculture is unimportant and does not require resources or a favorable policy environment, because its relative share in the economy declines.

Timmer (1998) suggested two mechanisms to explain why agricultural growth leads to agricultural decline. The first mechanism is stated in *Engle's Law*: because growth is led by demand patterns in market economies, a less-than-unitary income elasticity for the products of the agricultural sector guarantees that the gross value of sales by farmers will grow less rapidly than gross domestic production. However, if agricultural output fails to grow rapidly enough, rising prices might actually garner farmers a high share of consumer expenditures, and this would keep agriculture in a position of relatively high importance. The second explanation posits that when the terms of trade are unfavorable for agriculture, farm productivity must rise through technical change in order to offset low international prices. The combination puts pressure on agricultural resources to move out of farming and into the more rapidly growing industrial sector of the economy. Even with substantial increases in agricultural productivity, the relative contribution of agriculture to GDP actually declined in many countries. An almost identical situation occurs, as experienced in many contemporary developed countries, even if the economy is biased in favor of agriculture through price-distorting policies (Anderson, 1987).

The share of labor in agriculture also falls, except in cases where agricultural labor productivity grows extremely slowly relative to labor productivity in other sectors. Where labor productivity is growing more rapidly for farm versus non-farm jobs, there is even more reason to expect agriculture to become a less important employer. The number of people employed on farms in industrialized countries decreased in absolute terms at a rate of 2.8% per year during the period 1960 to 1980 (OECD, 1983; Mellor 1998).

Agriculture cannot therefore provide high overall growth rates in output and employment. Rather, its major indirect effects come mainly from the expenditure of increased agricultural income on non-agricultural goods and services, in turn creating not only additional output in those sectors but also additional employment. Farmers, meanwhile, do not necessarily lose out as a result of such an economic transition, as they benefit from the fruits of economic growth either through new technologies, which raise the productivity of farm labor and land, and/or through household members taking up better-paying jobs in other sectors (Schultz, 1964; Kuznets, 1966; Johnson, 1973). Such intersectoral movements of resources may well be painful in societies that have a less developed market infrastructure and mechanisms to cushion the effects of the adjustment process are likely to be country-specific.

The question is, would it not be costly for developing countries if policy-makers did not support (give priority to) investment in the agricultural sector? In addition, can the economy grow sustainably without agricultural sector growth? In the poorest countries, where 49% to 50% of GDP, 70% to 80% of the work force, and 70% to 90% of foreign exchange earnings are accounted for by agriculture, it is impossible to sustain broad-based economic growth without the active participation of the rural economy. Nicholls (1964) stated:

> *"Economic development will, of course, inevitably involve industrialization...this can be expected to follow, almost automatically, upon the growth of food surplus of the agricultural sector.... Once this is recognized, the efforts of*

underdeveloped countries could be concentrated far more than they are at present on tackling the problem of how to raise productivity on the land, as a prior condition of economic development."

Nicholls argued that until the underdeveloped countries achieved and sustained (either through domestic production or imports) a reliable food surplus, they would not have fulfilled the fundamental precondition for economic development.

2.2 The Role of Agriculture in Economic Development: Theoretical Framework of Intersectoral Linkages

Studies by the World Bank (1982) showed the continuing importance of agriculture in the economies of the developing countries. It found that there is a strong association between growth in agriculture and growth of GDP, except in oil or minerals-based economies. The parallels between agriculture and GDP growth suggest that the factors that affect agricultural performance may be linked to economy-wide social and economic policies. Expanding agricultural production through technological change and trade creates important demand for the outputs of other sectors, notably fertilizer, transportation, commercial services, and construction. At the same time, agricultural households are often the basic market for a wide range of consumer goods that loom large in the early stages of industrial development, such as textiles and clothing, processed food, kerosene and vegetable oils, and construction materials for home improvements. The failure of development strategies to emphasize agriculture in either capital allocation or public policy results in occasional crises in domestic food supply and/or large food imports. Ultimately, this hinders the process of industrialization. The significant contribution of agriculture to the overall economy is broadly explained in the context of traditional market-mediated linkages, referred to as *Lewis linkages of factor markets and Johnston-Mellor linkages of product markets* (Figure 2.1).

2.2.1 Factor Market Linkages

The Lewis linkages operate through factor markets that transfer labor and capital from agriculture to industry. In another seminal article, "*Industrialization is Dependent upon Agricultural Improvement*", Lewis (1979) observed that it is not profitable to produce a growing volume of manufactures unless agricultural production is growing simultaneously, and he underlined that the revolution in the two sectors should always go together. History has shown that higher productivity in agriculture provides labor and capital for the expanding industrial sector, to the mutual benefit of both sectors. Particularly important for the transformation are actions that promote intersectoral resource flows from agriculture to other sectors of the economy while still maintaining the profitability of agriculture. Two of the most important transfer mechanisms are voluntary investment of physical and human capital by farmers and by proprietors in non-agricultural sectors, and "invisible transfers" brought about by lower food prices (Staatz, 1998). Increased food marketing should also help depress food prices, thereby encouraging employment in other sectors by making labor cheaper relative to the goods and services it produces.

2.2.2 Product Market Linkages

In addition to the factor market linkages, Johnston, et al (1961) identified a broader set of linkages between agriculture and growth in the non-agricultural sector, which tend to be mediated by product markets that become progressively more efficient during the course of economic development. Contributions through the latter linkages are manifold. These are:

- Food for the industrial work force (thus avoiding deterioration of the terms of trade for industry);
- Raw materials for agro-processing industries;
- Markets for industrial output, especially for low-quality goods produced by domestic factories as part of a learning process but which cannot compete in export markets;
- Export earnings that pay for imported capital equipment and intermediate inputs.

2.2.2.1 Food Market Linkages

For enormous reasons, it is impractical to achieve at sustainable economic growth without first solving the problem of food shortage (Timmer, 1992; Delgado et al., 1994). The primary source of food for a society in general including workers in an industry is the agricultural sector. A well developed agriculture can provide workers in industrial sector with sustainable food supply at low (but profitable for producers) and stable price. What if the food sector is ignored and the industrial sector is focused upon? The consequence is that employment in the latter sector increases with increased income, the largest proportion of which is spent on food. This will cause the price of food to rise, squeezing the real income of laborers downwards, close to the previous level, and reducing the incentive to work. This in turn places upward pressure on wages, and ends up reducing employment and/or industrial profit. Therefore, the capacity of the industrial sector to produce a growing volume of manufactures could be restricted unless agricultural production is growing simultaneously. This is also the reason why the industrial and agrarian revolutions always should go together, and why economies in which agriculture is stagnant do not show industrial development (Lewis, 1978). Above all, increased food security of the poor can contribute substantially to short and long-run economic growth. Securing food encourages farmers to invest more in innovations and technologies that increase farm productivity (Zeller, et al, 1997). An important component of this investment is human capital, which is crucial in the overall process of economic development.

The dilemma in developing a food strategy is how much emphasis should be given to food production. There appears to be no concrete answer to this question, as it depends on the country's resource potential to produce and the demand it has for food. At any rate, over-investment in local food self-sufficiency in resource-poor zones may be a very costly way of promoting food security (alleviating widespread chronic hunger) that drastically retards economic growth (Mellor, 1998).

Figure 2.1: Backward and Forward Agriculture-Industry Linkages

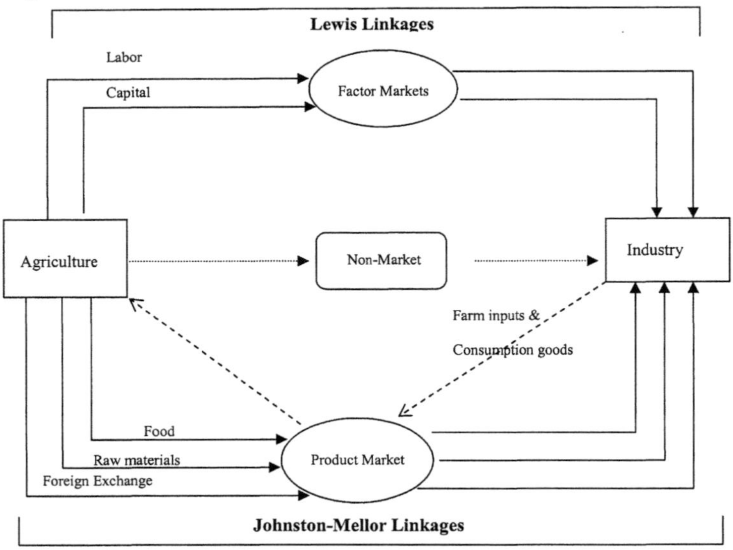

Source: Own illustration

2.2.2.2 Raw Materials Linkage

The importance of agricultural raw materials for industry is extensively documented in Ohkawa (1964). Once food supply is secured, the agricultural market surplus is available as input for domestic agro-processing industries, which is used for production of consumption or intermediate goods. A domestic supply of raw materials stabilizes fluctuations in the availability and prices of imported inputs. Moreover, it saves foreign exchange. Thus, it promotes self-sustained development of the industrial sector, particularly in the early stages.

2.2.2.3 Demand Linkages for Products of the Non-agricultural Sector

The third important role that agricultural sector can play is providing effective demand for products of the industrial sector (backward linkages). This includes rural agriculture-based households' demand for consumption goods and farm inputs produced by the industrial sector. The expenditure pattern from the net additions to income arising from accelerated growth of agriculture creates demand for a wide range of goods and services with a high employment content, much of the production of which must be broadly diffused in rural areas (e.g., in major market towns). In fact, this can occur only through accelerated growth in employment (more precisely, increased demand for labor), which is facilitated by the indirect effects of agricultural growth itself. Thus, agriculture offers a potential for rapid growth in domestic demand for labor-intensive goods and services. Research work by Chuta and Carl (1990) supports this, and Haggblade et al. (1989) found that rural households' expenditure elasticity is highest for local non-food goods and services. However, Eleni et al. (1999) assert that the magnitude of effective demand for non-agricultural products depends on the income of rural farm households that can be used to purchase farm inputs and non-farm consumption goods and services.

2.2.2.4 Linkages in Foreign Exchange Earnings (A Comparative Advantage Approach)

As long as a country is not industrialized, export earnings largely come from primary products generated by the agricultural sector. This income, combined with domestic farm sales income, is the important part of the net capital outflow from agriculture that can be transferred or used to finance imports of industrial equipment and other investment outside agriculture, such as investment in infrastructure development (Mellor and Lele, 1973; Wade, 1990). Accelerating the growth of agricultural production therefore enlarges the volume of exports and increases foreign exchange earnings.

2.2.3 Non-market Linkages

The other contributions of agriculture arise from non-market mechanisms, i.e., the efficiency with which rural households allocate resources at their disposal and the low opportunity cost of much of the household labor. Making more resources available to these households, in the form of higher incomes or new technologies, raises factor productivity (efficiency) for the entire economy (Timmer, 1995). Rural savings are seldom captured, if at all, by the national income account, because financial intermediaries are few or absent altogether. These savings are held either as liquid but non-productive assets, such as gold or jewelry, or they are invested in non-liquid but productive assets, such as livestock, farm implements, etc. (Morduch, 1991). If the rural economy is transformed from its traditional, risky nature to a dynamic one that is attractive for investment, then these savings can be channeled directly into more productive investments on the farm or in the local economy, even where financial intermediaries are totally absent (Birdsall et al., 1995; Timmer 1995). Figure (2.2) provides a visual summary of the mechanisms, sectoral linkages and process of agrarian structural transformation

Figure 2.2: The Process of Structural Transformation of the Agrarian Economy

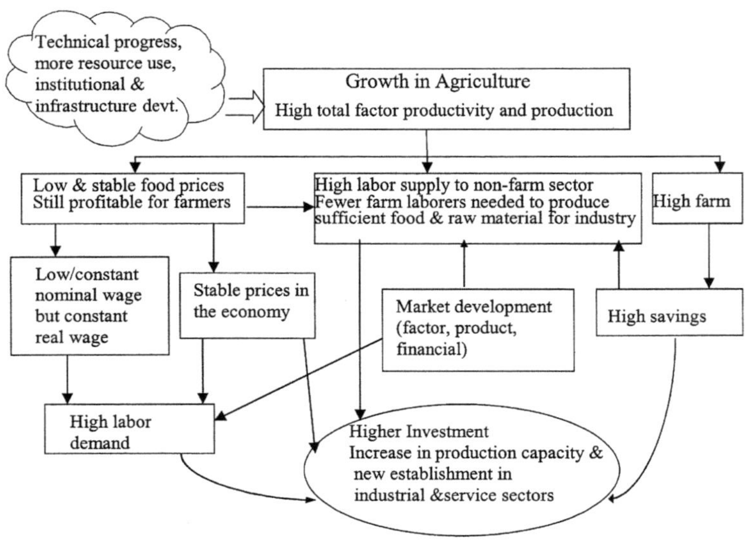

Source: Own demonstration

In all these processes of intersectoral linkages, it is worth noting the decisive importance of the need to develop efficient markets for all inputs, product and financial markets. For instance, the transfer of labor depends on labor supply, which is a function of the labor market and the food market. Capital needed for agricultural growth and for transferring to the industrial sector can be generated in agriculture itself in response to a technology-induced high rate of return. The implication is that for the process of structural transformation to go forward, the economy must develop low-cost means of exchange; market mechanisms should work well to raise the savings rate, and they should be informative to farmers to encourage them to invest in efficient production (or in the industrial sector). High transaction costs in an economy can choke off structural transformation by making it too costly for people to rely on the specialization and exchange necessary to take advantage of the new technologies in the system. This requires an ongoing stream of technological and institutional changes throughout the system.

In conclusion, the route to rapid economic growth leads from the rapid increases in agricultural output to food security and from those through a maze of linkages to a rapidly growing industrial economy. These linkages are the most important connection between agriculture and economic growth that a government should strengthen by investing in agricultural development, accompanied by conducive economic policies, institutions and politics for producers and consumers (Barrett and Carter, 1994).

2.3 Accelerating Agricultural Growth: Precondition for Structural Transformation of an Economy

Once the role of agriculture in economic development is appreciated, it becomes important to discuss the ways and means of promoting agricultural growth. The most critical determinants of growth in agriculture are included in this section, with a brief explanation of the concept of agricultural transformation.

2.3.1 Agricultural Transformation: Concepts and Conditions

Agricultural transformation is also referred to as agricultural modernization. It means the gradual and sustainable transition of individual farms from subsistence-oriented to diversified and then towards more specialized market-oriented production. Todaro (1989) summarized the characteristics of each stage of farm evolution in the transition process. In subsistence agriculture, productivity is low and capital investment is minimal, while land and labor are the principal factors of production, and simple traditional methods and tools are used. Family members manage their farm without hired labor. Labor is seasonally underemployed, a phenomenon also referred to as disguised unemployment. Most output is for the household's own consumption, and staple foods dominate production. Subsistence agriculture is highly risky, and the main motivating force in peasant farmers' lives may to maximize their family's chances of survival, technically maximizing utility, but not profit.

Diversified agriculture refers to the stage where part of the produce is grown for own consumption and part for sale to the commercial sector. Production is no longer dominated by staple food crops, but includes basic animal husbandry and cash crops such as fruits, vegetables, coffee, tea, etc., which take up the slack time of the year when disguised unemployment is usually prevalent. The transformation eventually leads to a final stage of transition in which highly productive and specialized modern farms produce exclusively for the commercial market. In this stage, many functions formerly performed on the farm, such as input production and output processing, are shifted to off-farm areas of the economy. Note, however, that all three types of farm may be found coexisting in almost all least developed countries (LDCs) at any given time. For a majority of developing countries, peasant subsistence and small-scale mixed family farms dominate the contemporary agricultural system.

Mellor (1998) noted some important conditions for agricultural transformation. From the supply side, the pace of agricultural growth must be accelerated despite the limitation in total area cultivated. From the demand side, domestic demand for agricultural output must grow rapidly despite inelasticity of demand. In addition, fundamental interrelated conditions and stages of evolution need to be fulfilled to foster structural transformation of a predominantly agrarian economy (Timmer, 1998; Eleni et al., 1999).

First, there must be investment in agriculture to substantially increase its productivity and then to create surplus. This expands the market share of agricultural output, thereby increasing farm cash income. This stage is about "getting agriculture moving". Second, the surplus generated in agriculture can be utilized to develop the non-agricultural sector, i.e., investments in agriculture begin to pay off, and the agricultural sector becomes the key contributor to the overall growth process through a combination of factors (Johnston and Mellor, 1961). Therefore, there must be a positive interaction between agriculture and industrial development. Rural factor and product markets must become better integrated with those in the rest of the economy. Efficient labor and credit markets are among the main requirements for linking the urban and the rural economies. This is because improved functioning of factor markets speeds up the process of extracting labor and capital from agriculture, where returns are low, and shifting them to industry or services with higher productivity. This results in an increase in non-farm employment relative to the growth of the total labor force. The progressive integration of the agricultural sector into the macro-economy via improved infrastructure and market equilibrium ultimately results in an agricultural sector that is increasingly integrated into the rest of the economy. Therefore, to begin with agriculture, expansion and/or intensification of agricultural resource use and enhancing factor productivity are the two classical methods of accelerating growth on the supply side. The other equally important factors in terms of accelerating the rate of transformation are availability of adequate effective demand and an appropriate institutional and policy environment. Figure (2.4) portrays the interrelated sources of agricultural growth.

2.3.2 The Role of Resource Base and Efficient Price: Classical and Neo-classical Approach

A more theoretical approach to development in the literature of neo-classical economics is that based on assumptions related to individual decision-making behavior and market functioning. First, individuals constantly and efficiently maximize their satisfaction as consumers and their profits as producers; and second, markets are free and prices are therefore flexible to adjust to equilibrium supply and demand. The importance for small and medium-sized farmers of price incentives that truly reflect international market conditions is emphasized. This often means less government intervention or allowing free market policies. The price paradigm presumes the transfer of income from a relatively small number of urban producers and consumers of importable goods to a relatively large number of rural producers. This, it is argued, is done by a combination of devaluation, tariff reform, and direct pricing policy action to shift the internal terms of trade in favor of agricultural producers.

Assuming constant returns to scale in each line of production, the theory explains how full utilization of resources and optimal allocation for various uses can maximize the aggregate value of production in the sense of "Pareto optimality". According to the theory, the level of national income is therefore determined by endowment and stock of productive resources such as land, labor and capital. The level of these resources subsequently determines the structure of output via the law of demand and supply.

The theory goes on to show how an increase in the stock of resources, especially capital and labor, leads to growth of aggregate output over time. The rate of capital accumulation (the real value of all physically productive capital goods and human resources) depends on the savings ratio (the proportion of income that is saved and invested), which in turn is mainly determined by the level of income. This is best described by the *Linear Stage Growth* model. Most attention is focused on mobilization of domestic and foreign savings to generate sufficient investment to accelerate economic growth. This can be illustrated by a simplified version of the Harrod-Domar growth model:

$$\Delta Y/Y = s/k$$

Where $\Delta Y/Y$ represents the rate of growth of GNP, s is the savings rate, and k represents the incremental capital output ratio ($k = \Delta K/\Delta Y$), which is the marginal productivity of capital. The Harrod-Domar equation states simply that the rate of growth of GNP is determined jointly by the national savings ratio, s, and the national capital/output ratio, with their respective directions of effect. The model assumes that savings equals investment, i.e., all savings will be invested in an economy. According to the model, to foster economic growth is simply to increase national savings and investment. Therefore, poverty in a country is explained by its small factor endowment relative to its population size, and growth in such countries is explained by an increase in their factor endowment, especially capital stock (Sundrum, 1983).

In this approach, expansion of the area cultivated or of grazing land represents a predominant means of increasing agricultural production. It implies the activity of clearing forest or jungle land for cultivation of annual and perennial crops. If soil conditions are found to be favorable, farmers tend to intensify their cultivation. Investment in mechanization, irrigation installation, and the use of chemical fertilizer and insecticide are recommended to increase land productivity. This model of agricultural development is also called the "frontier" model (Ruttan, 1998).

However, such an approach leads to a vicious cycle: the Less Developed Countries (LDCs) save little because they are poor, and they remain poor because they save little and cannot accumulate capital stock. If successful at all in terms of increasing incomes and reducing poverty in LDCs, the process will take an unduly long time. Todaro (2003) argued that although growth in productive resources, both in quality and quantity, is a means of accelerating the growth of the national economy in the long run, this is not an economic law, nor is it even a necessary condition for short-run growth. Sundrum (1983) underlined that savings and investment, on the other hand, are a necessary condition, but not in themselves a sufficient one.

Moreover, the approach is not completely satisfactory as an explanation of sources of growth of agriculture in LDCs for three main reasons. First, capital accumulation does not necessarily occur, as resources are still underutilized. Many contemporary developing countries operate even with underutilized existing capital and/or land resources and unemployed labor; secondly, the market is not perfect as it is in developed economies, and finally, there is limited potential for further expansion in the use of resources such as land. Increasing pressures on land may ultimately result in declining crop yields and environmental degradation. Moreover, increased output that can be achieved only via a corresponding increase in input intensity (which implies rising costs) must be offset by rising product prices. This is evidently not sustainable, and it is not socially acceptable. Hence, the structuralists argue that agricultural transformation must be derived from technological change complemented with infrastructure development and institutional innovation.

2.3.3 Technical Progress or Productivity Growth: the Structuralist Approach

Technical progress refers to technological progress and increased efficiency in undertaking activities. It can originate from embodied or/and disembodied technical changes. The embodied technical change includes new and improved machinery or material that can be resulted from new knowledge embodied in its development (constitutes technological change), and the disembodied technical change is changes in the knowledge, techniques or management practices or methods of organization to accomplish traditional tasks (constitutes efficiency changes). Both lead to improvement in productivity of resources, and thus produce changes in production coefficients. In agriculture, these refer to the acquisition of new/improved seeds, chemicals, equipment and the acquisition of new production or husbandry management skills. It also includes the development of new skills in dealing with both natural and market institutions linking agriculture with the non-agricultural sectors (Ruttan, 1998).

Technologies are developed to facilitate the substitution of relatively more abundant (hence cheap) factors for relatively scarce (hence expensive) factors of production. Major sources of technological innovations that can increase farm yields include mechanization, water control (irrigation), and biological and chemical innovations. Mechanizing agriculture is a means of replacing human labor (labor-saving technology e.g., tractor) where land is extensively cultivated but labor is scarce. Biological technology refers to genetic improvements in plants (e.g., hybrid seeds) and animals, and chemical innovations include fertilizer, pesticides, insecticides and herbicides. Biological and chemical innovations are called for where land is scarce, and hence they are said to be land-augmenting technologies that increase yield per hectare, but they only indirectly increase output per worker. They are also scale-neutral (i.e., theoretically they can be applied equally effectively on large and small farms), and do not necessarily require large capital inputs or mechanized equipment. Biological and chemical technologies are therefore particularly well suited to tropical and subtropical regions and offer enormous potential for raising agricultural output in the LDCs.

Technological change solves major problems of agricultural growth and allows low-income countries to use the most powerful element of growth. Schultz (1964) asserted that farmers and herders have learned how to allocate efficiently the factors of production available to them through a long process of experimentation. He then concluded that, given their existing technologies, these farmers are "efficient but poor" because in most poor countries there are only limited technical and economic opportunities to which they can respond. This implies that no appreciable increase in agricultural production is to be made by reallocating the factors at the disposal of farmers who are bound by traditional agriculture. Therefore, a major increase in per capita agricultural output would be achieved if farmers gained access to new and more productive factors of production (i.e. new agricultural technologies) and the skills needed to exploit them.

Technological progress consists of four stages (Nabseth and Ray, 1974). These are: i) discovery of new products or processes, i.e. inventions; ii) development of inventions into commercially feasible and profitable techniques, i.e. innovations; iii) modification of innovations in the course of actual production, i.e. the learning process and; iv) the spread of profitable innovations among farm households and private firms in a region or among countries, i.e. diffusion. This progress occurs either exogenously or endogenously under the influence of market forces. In the earliest theories of development, technological progress was assumed to occur only exogenously. Recently, however, it has been recognized that technological progress is rather an endogenous process, particularly in a market economy. Ruttan and Hayami (1998) also hypothesized that technical change is guided along an efficient path by price signals in the market, if the prices reflect changes in the demand and supply of products and factors and if there is effective interaction among farmers, public research/extension institutions, and private agricultural supply firms. Farmers are induced by shifts in relative prices to search for technical alternatives that save the increasingly scarce factors of production. Moreover, they press the public research institutions to develop new technologies and press input supply firms to offer modern technical inputs as a substitute for the scarce factor. Generally, changes in relative prices in any sector of the economy act to induce innovative activity, not only by private producers but also by scientists in public institutions, in order to reduce the constraints imposed by those factors of production that are relatively scarce. Therefore, policy and institutional changes should accelerate growth in agriculture and thereby the process of structural transformation.

2.3.4 Policy and Institutional Change (Promoting Efficiency)

The other side of technical progress is improving technical and allocative efficiency. Improved institutions and policies are necessary to increase resource use efficiency, promote effective competition and social and economic mobility, permit greater equality of opportunities and ultimately lead to higher productivity (Ruttan and Hayami, 1998). Therefore, economic growth results not only from growth in the quantity and quality of resources and improved technologies, but also from social and political structures that are conducive to such changes and enhance efficient production. Growth demands a stable but flexible social and political framework that is capable of accommodating and even encouraging rapid institutional and structural change.

A developing country that fails to evolve a capacity for technical and institutional innovation in agriculture consistent with its resources and cultural endowments will suffer from an inability to take full advantage of advances in biological, chemical and mechanical technologies suited to the agricultural system (Ruttan, 1998; Schultz 1978; and Ohkawa, et al, 1969). Therefore, parallel reforms and development of institutions related to land tenure, rural finance, education, research and extension services, factor and product markets, communications, infrastructure and system of administration and planning are imperative.

In an agricultural system characterized by static technology, only few gains can be realized from education in rural areas, because they have lived and worked with the same resources for generations and know how to use these resources efficiently (Schultz, 1964 and 1978). Conversely, the effects of education are particularly important during periods of transition, when a nation introduces new technologies into the economic system in order to enable them to cope with the dynamic changes in optimum input and output mix in the process of economic growth.

The level of market development and the degree of market integration affects the speed of economic development through its direct impact on efficient use of productive resources. Crucially important in the agricultural development strategy is that farm households face a cash constraint as regards using improved purchased inputs. Zeller et al., (1997) documented the role of access to credit and savings in this context. They provide capital to finance farm inputs and labor, and enable adoption of profitable but risky technologies. Moreover, credit alleviates food insecurity. Savers hold financial rather than real assets, which can be reinvested to achieve a higher real return through financial institutions. Saving and investment are the major route for intersectoral transfers to facilitate the generation of non-farm activities that expand employment and income, which are essential complements to improving agricultural production. It is therefore necessary to develop credit institutions that assist farmers by providing credit for the use of such inputs.

Timmer (1998) stated that a significant share of a country's inevitable resources (surplus) might well be extracted from agriculture in the early stages, as the rest of the economy is so small. In this case, direct and indirect taxation of agriculture is the only significant source of government revenue. These resources need to be allocated to public investment in research and infrastructure. In other words, in order to build a dynamic agricultural sector, these resources should be devoted to agriculture itself. In this regard, Hayami and Ruttan (1985) remarked that caution must be exercised when adopting a national policy that impacts on the allocation of resources between technical and institutional innovation, and between the agricultural and the industrial sectors. It must be consistent with national or regional resource endowments if it is to lead to an efficient growth path. It involves an adaptive response on the part of cultural, political, and economic institutions in order to realize the growth potential opened up by new technological alternatives.

Figure 2.3: Effect of Resource Use and Technical Progress in Agriculture on an Economy

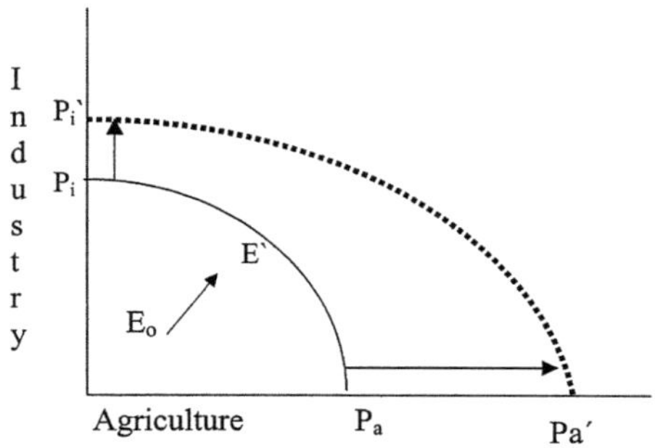

Source: Adapted from Todaro, Michael (2003)

A production possibility frontier (PPF) can be used to summarize the role of resource expansion, technological progress and efficiency improvement in the production system or in an economy. Note that the PPF portrays the maximum attainable output combination of any two commodities with a given technology and amount of physical and human resources, operating with full utilization of resources. See a two-dimensional presentation of PPF in Figure 2.3. Both expansion in resource use and technological advancement shift the PPF outwards while improvements in resource use (efficiency) push the combination of commodities produced upward to the PPF. To show simple cases of changes, suppose that P_i-P_a is the initial PPF before interventions for a combination of agricultural and industrial products in an economy. Assuming that the technology remains the same, new investments in agriculture will increase the quantity of factors used in the sector. This will cause the entire PPF to shift outwards from P_i-P_a to P_i`-$P_a$`. Since growth in agricultural production causes growth in the industrial sector through forward sectoral linkages, production of goods in both sectors increases. However, the change will be more pronounced in agriculture in the early stages of development. If resources are initially underutilized, then the combination of production is indicated inside the PPF, say, at point E_o. The distance from E_o to the PPF depends on the level of resource use efficiency; in other words, the closer E_o is to the PPF, the higher the efficiency of resource use. Better utilization of idle resources can therefore raise output levels substantially (move from E_o to E`). Technological changes such as the use of high yielding varieties have similar effects. Similarly, technological progress or investment in the use of more resources in the industrial sector shifts the curve outwards along both axes as long as the industrial sector produces goods and services that can be used by the agricultural sector (backward linkages).

2.3.5 The Importance of Effective Demand on Agricultural Development

Solving the problem of poverty in low-income countries requires rapid growth in output, income, and employment. According to the production function approach of classical and neo-classical economics, the aggregate level of production in a country is determined by its productive capacity (potential production) of resources. The role of demand was neglected in the belief that supply creates its own demand; i.e. via flexible prices that speedily adjust supply and demand in each market to the equilibrium point, as explained by such devices as Marshall's assumption of instantaneous price movement, Walras's auction and Edgeworth's contract assumption. However, there have been many examples of increases in output due to an increase in demand in developed countries and even in LDCs (Sundrum, 1983). The most important point is that desired purchases or sales in one market depend on the nature of disequilibria in other markets. For example, the demand for labor by a firm depends on the market demand for its products; in other words, the higher the market demands for the product of the firm, the higher the firm's demand for labor. This implies that producers invest if there is a market for their products.

However, where is the effective demand for massive increases in output to come from? Although foreign demand requiring open trading regimes and large growth in exports is currently much talked about, domestic demand has traditionally been most important in the initial stages of accelerated economic growth. Growth in domestic demand provides the basic markets both for increasing agriculture and for the activities that create rapid growth in employment. Thus, to realize the full potential of technological change, there must be sufficient demand particularly for food (elasticities at low levels of income are usually high), and for raw materials produced by the agricultural sector.

Figure 2.4: Sources of Agricultural Transformation

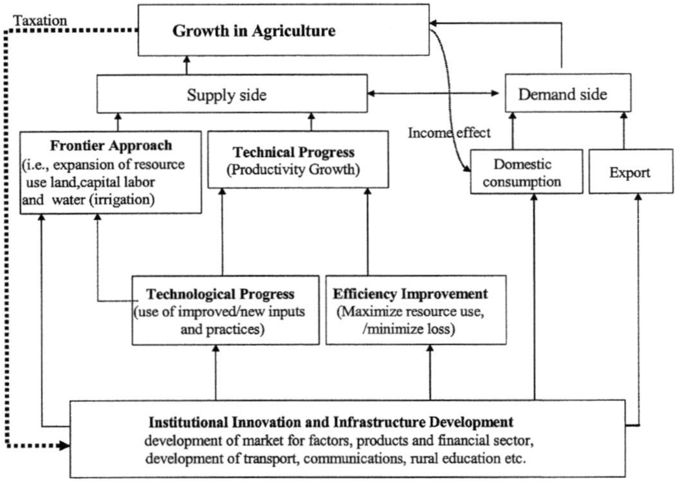

Source: Own illustration

Kick-starting domestic demand calls for increased productivity in the large agricultural sector itself through attractive prices that encourage investment in agriculture and efficient resource use (Mellor, 1998). Increased food demand can also be achieved through simultaneous accelerated growth in employment, which in low-income countries is efficiently transferred to an increase in effective demand for food. Employment creation and public sector spending in infrastructure and agriculture lend more to multiplier expansion, as more domestic inputs are involved than imported raw materials, capital goods, equipment and technology. Expansion of rural incomes boosts domestic demand as more local goods and services are demanded in the first instance as the rural population sees incomes rise. Low (2001) found that in many large agricultural rural populations in Asia, for example in China, India and Indonesia, the marginal propensity to consume is high and the marginal propensity to import is low. The increase in demand for agricultural raw materials can be also met by establishing domestic resource-oriented manufacturing and processing industries. Indeed, orienting an economy towards domestic demand to stimulate growth does not imply a closed economy and distorted prices. Conversely, if technological change moves ahead of employment, food prices (prices of agricultural output) decline, giving rise to exports even though the local need for food has not been satisfied. Otherwise, technological change merely results in a fall in prices that ultimately handicaps rapid growth in agriculture and limits the benefits associated with it, because of the "agricultural treadmill effect".[3]

2.4 Does Investment in Agriculture Continue to Pay in the World of Globalization? Concepts and Arguments

Globalization essentially refers to the expansion of international trade and financial market because of more open national economies. It is driven by technological revolutions in the transportation, communications and information sectors. These three revolutions have significantly reduced the cost of economic transactions and significantly increased the scope of markets and international trade. Although the benefit of globalization to the poor is debatable, the benefit from international trade is widely accepted, and literature on this is robust. Globalization is seen by some as the way of the future; it brings unprecedented prosperity to everyone, everywhere. Others fault globalization as the source of untold problems, from destruction of native culture to increasing poverty and misery life. That implies globalization has meant different things in different places. In this regard a close observation by Stiglitz (2002) is worth mentioning. His primary concern is that market is not perfect, and market failure is inevitable. Thus, it becomes a crucial importance that governments have to intervene in the market operation in a systematic manner, the argument which appreciates *Keynes* theory of development. According to Stiglitz, globalization has immense benefits and costs, but the net benefit depends on how it is managed. The countries that have managed globalization on their own, such as those in East Asia, have by and large, ensured that they reaped huge benefits and those benefits were equitably shared. Their growth has been based on exports and by closing the technological gap and creating their own enterprises. Each of the successful globalizing countries determined its own pace of change; each rejected the basic argument for minimalist role for government and rapid privatization and liberalization. In Korea and Taiwan, for instance, government took active role in managing the economy. Financial markets were highly regulated because rapid liberalization of financial and capital market can easily result in high economic volatility that impedes growth and increase poverty.

By contrast, the countries that have had globalization managed for them by the IMF and the World Bank have not done so well, e.g., many sub-Saharan countries. According to Schuh (2002), in such countries globalization makes a country more vulnerable to changes in the international economy and become

[3] The "treadmill effect" refers to the case in which the successive widespread adoption of new agricultural technologies results in an upward shift in aggregate supply of that commodity and a decline in its price. Because demand is inelastic, gross returns to producers fall (Eleni et al., 2002).

divorced from the control of the national economic power (a nation becomes likely to lose its sovereignty). Stiglitz (2002) also strongly argues that the trade liberalization agenda has been set by special interests in the North. Globalization is condemned primarily for exacerbating income inequalities. Gains from technological breakthroughs have benefited developed countries disproportionately, whereas the developing (poor) countries have not shared many of the benefits. The United States and Europe gained enormously, but sub-Saharan Africa, the poorest region of the world, lost by about 2% because of unfavorable terms-of-trade effects. The trade negotiations of the Uruguay Round in 1994 has opened markets to manufactured goods produced by the industrialized countries but has not for agricultural goods in which poor countries often have a comparative advantage. Nor has the trade agreements eliminated subsidies to agriculture in the developed countries that make it so hard for developing countries so compete.

Applied unevenly to favor urban industrial rather than rural agricultural sectors, globalization has accentuated income and employment insecurity, especially among the vulnerable less educated, less skilled and minority groups and widened the digital divide (Low, 2001). In rural communities, the most important problem is that the supply and demand structure of agriculture (which is inelastic at least in the short-run) could create circumstances in which labor flows out of agriculture at an early stage of development. The drain of human capital from the rural areas could weaken the basis for further economic development of the sector. The globalization managed by the IMF and the World Bank has resulted in rather than moving workers from low-productivity jobs to high-productivity ones, moves them from low-productivity jobs to unemployment. Rather than enhanced growth, the effect is increased poverty and vulnerability to risks of hunger and famine (Stiglitz, 2002).

On the other hand, the ever-faster international transfer of information, ideas, capital, skill, and new technologies (such as new genetically engineered crop seed, pesticide and fertilizer industries) reinforces opportunities for reform of the agricultural sector in the environment of globalization. This indicates that the sources of economic growth will depend increasingly on improved access to the knowledge systems of the wider world, as embodied in new technologies, management practices, and institutions. Analysis of trade highlights the role that exports can play in stimulating increased intensity of agriculture through a change in output composition towards higher-value crops, particularly in the interim while domestic income growth is gathering steam. It also provides healthy competition that reduces the cost of production to increase productivity, which makes a net addition to national income (Mellor, 1995). High savings and prudent spending are also virtues worth consideration in designing policies to stimulate the rural poor toward globalization (Low, 2001).

Nowadays, however, the debate is not so much about accepting or rejecting globalization, as it rather assumed to be inevitable within the growing interdependence of the world's economies and markets. A more important goal is to ensure that positive attributes of globalization for the poor are maximized and that negatives ones are minimized. This holds for both the developed and developing countries. Bale (2001) proposed that most priority areas of policy interventions should do this. Developing countries should first foster broad-based rural growth that rests on policies working toward improving agricultural productivity and competitiveness through investment in science and technology and promotion of institutions and rural infrastructure. Improving social well-being, for example by reducing the risk of and susceptibility of the rural poor to food and health insecurity caused by internal and external shocks such as climate and price fluctuations, is the second key area of policy intervention. The third important point is to enhance efficient use of natural resources for sustainable development. Schuh (2002) suggests community integration, in which case neighboring communities agree to integrate their existing institutions, thus making more efficient use of their resources and reducing costs. In this respect, experiences in Asia have also shown that globalization has been favorable but with careful government policy control. It is also learnt that raising agricultural productivity remains a long-standing imperative when supplemented with development of information technology in production, marketing and distribution in the rural sector.

Clearly, the industrialized countries and the development community can provide much support and encourage rural development by cushioning the negative effects of globalization in a number of ways. Bale (2001) suggested first increasing the amount of aid and making its composition development-oriented to improve the efficacy of donations. The second is revising the macroeconomic policy of the industrial countries to induce stability and greater integration of developing countries into the global economy. For instance, trade polices are seen as biased against poor countries: vigorous enforcement of anti-dumping and phytosanitary regulations is seen as inappropriate when applied to developing countries. Moreover, the developed countries abundantly subsidize their farmers. This pushes down world prices for agricultural products and then deprives poor farmers in third countries of markets. Hence, unless the developed countries reconsider and resolve such problems, globalization will remain one-sided, functioning at the expense of the world's poor. As stated by Hoekman and Anderson (2000), therefore, the "Uruguay Round Agreement" on agriculture requires practical reductions in farm export and production subsidies and an increase in import market access. Based on multilateral agreements on agriculture, the focus should be on reducing agricultural protection (e.g., escalating tariff bindings in manufacturing goods and services) in industrialized countries to give farmers in developing countries better access to export markets.

In general, no single policy can ensure only positive outcomes for the rural poor and widely share the benefits of globalization. However, a comprehensive and holistic view of rural development is needed to take into account the many, interactive variables that can be marshaled to reduce poverty. A cooperative and coordinated approach by governments of developing countries and sympathetic supporters from wealthy countries will enable the poor to benefit significantly from globalization and reduce poverty. Conclusively, beginning from the early development breakthroughs of the 16th and 18th centuries to the present-day world of globalization, development in agriculture and agricultural productivity remains the principal source of overall economic development.

2.5 Empirical Evidence

Evidence has shown that there are different paths of growth and development, which depend on the socioeconomic and political situations of a particular nation. Nevertheless, the majority of developed and rapidly developing countries do have many things in common in their path to growth and structural transformation. Accelerating agricultural growth as an initial and fundamental step has been the principal strategy. Experiences – both successes and failures – of development strategies are documented in this subsection for distinctly different classes of nations; most developed (western) countries, rapidly growing (Asian) countries and lagging sub-Saharan African countries.

2.5.1 Important Development Histories of the Contemporary Developed Countries

Only small groups of countries broke away from the traditional system with remarkable development in the 18th century. The first major breakthrough from the agrarian, handicraft economy to one dominated by industry and mechanization occurred in Britain in the middle of the 18th century. Turner (1998) emphasized that this transformation was important in explaining early modern economic growth and laid the basis for the industrial revolution. Britain's industrial revolution occurred with dramatic changes in industrial technology, predominantly in manufacturing industry, and especially in the field of textiles, ironworking technology and steam engines, which Kuznets (1966) described as "modern economic growth". It then spread to other countries of Europe and to European settlements overseas in North America and Australia, and then to Japan in 1870s and Russia in the period from 1890-1914. In these countries, a sharp discontinuity in economic trends occurred, which Rostow (1960) described as a "take-off stage". Such a change in economic growth was initially brought about by a process of diffusion of new technologies and associated institutions. Governments that responded quickly and attempted to ensure greater scope for the efforts of individuals adopting the new technology for industrial production, and to provide the infrastructure and institutions needed for that purpose, established modern economic growth without much delay (Sundrum, 1983).

A particularly significant factor for the speed of diffusion of modern economic growth was the movement of people (learning from one another), mainly out of Western Europe to North America and Australia, and from Japan to Europe, rather than the movement of goods or capital (Arndt, 1981). In the early stages, migration of skilled workers from the more advanced European countries to the less advanced countries with relatively simple and easily diffused borrowed technologies accelerated the process of adoption. The application of new technologies was accompanied by a higher rate of investment. Capital accumulation and profitability of the technologies, therefore, were the driving force of the system. Under the pressure of technological progress, there were also major changes in the economic institutions of society (Sundrum, 1983). In general, modern economic growth in such countries was achieved by systematic imitation of one another. In countries that had an adequate level of agricultural productivity, growth spread under the influence of market forces, particularly in the early stages.

The transformation of the British economic structure required a corresponding increase in agricultural labor productivity (between 1600 and 1750, it grew at 79% despite the then rapid population growth in the country). Turner (1998) mentioned that three factors contributed to the growth in agricultural productivity in Britain in the early period of its development. The first is the advancement of technology. The second factor is the development of *quasi-proprietorial land tenure*, which provided an incentive for productivity growth. It guaranteed tenants the increased income from the growth in productivity for the duration of the lease (the carrot) while also requiring that the tenants save or otherwise finance the fine for renewal of the lease (the stick). The third factor is the rapid development of urban centers, which induced efficient use of resources by farmers through labor market linkages.

Kuznets (1966) found that in the United Kingdom, the United States, Japan, and the USSR, the share of agriculture in the total labor force and the share of agricultural output in the economy generally declined. In addition, the share of the labor force always fell more rapidly; this was because labor productivity in agriculture rose more quickly than in the economy as a whole when measured over the longer period. This resulted in sustainable economic growth and caused substantial changes in the structure of the economy. In the USA up to 1920, land expansion was the primary cause of supply growth. Between 1914 and 1956, total farm output increased by 90%, resulting almost entirely from technological advances due to biological and chemical breakthroughs. Similarly, the Asian "Green Revolution" was primarily a story of output growth resulting from technological advances.

Spreading from Britain, further diffusion of modern economic growth culminated in dividing the world into developed and underdeveloped countries, a division that persists to this day. Many explanations have been put forward as to why this type of growth did not diffuse to the present-day LDCs. Three most notable reasons are: level of initial income, choice of trade option and the low level of agricultural productivity (Sundrum, 1983). Rostow (1975) attributes the failure of modern economic growth in LDCs to the absence of certain prerequisites, including a higher level of income to begin with, a higher rate of savings, a more favorable attitude to science and greater awareness of fertility control. The second reason was what Lewis (1978) termed the 'trade option'. Contemporary LDCs might have preferred to produce agricultural products for export. This was likely to occur because of the revolutionary changes in ocean transport, which halved transportation costs, and the growing demand for tropical products in developed countries resulting from the growth of their incomes and industrial production. The other important reason was the case put forward by the "dependency school", notably by Frank (1967). He argued that colonial governments, acting in the interest of the metropolitan countries, dictated the pattern of development in the tropical countries. This was partly the result of the free trade regimes established by the colonial governments, especially between each metropolitan country and its tropical colonies for the specific purpose of promoting exports of primary products (Birnberg and Resnick, 1975). This led some writers to argue that economic underdevelopment in LDCs was due to their economic relationship with their colonizers in the developed countries.

A more significant factor than those mentioned above, however, was the level of agricultural productivity. First, low agricultural productivity in the tropics inhibited the diffusion of modern economic growth because it limited the market for industrial goods. Secondly the high productivity of labor in agriculture in the temperate countries, which could support a European wage level, attracted migration from poor and overpopulated tropical countries where agricultural productivity was much lower (Lewis, 1978). For instance, in the case of Japan, there was a rapid increase of agricultural productivity on the one hand, and, on the other, a substantial flow of Japanese people to Europe who brought back the new techniques to Japan (Sundrum, 1983). After these countries had achieved industrialization of their economy, agriculture continued to play a major role in supplying food. Thus, governments still continue to protect agriculture from foreign competition through subsidies and import taxes to stabilize the food sector.

2.5.2 Agricultural Growth and Economic Transformation in Countries Currently Undergoing Rapid Development

In the period from the 1950s to the 1980s, the performance of the agricultural sector in developing countries was relatively poor due to government neglect of the sector and a bias towards investment in the urban industrial economy that can be traced largely to a misplaced emphasis on rapid industrialization via import substitution and exchange rate overvaluation. With due consideration of agriculture, remarkable efforts by international agricultural research and extension organizations resulted in new, high-yielding crop varieties, and new water and soil conservation techniques. This research was supported by investments in rural infrastructure and irrigation, increased inputs such as fertilizer and pesticides, and resulted in the implementation of policies that regulated agricultural trade, commerce and credit. Sustainable incentives for private investment in agriculture were implemented in many countries, and the dynamic demographic and labor shift culminated in structural transformation. This economic transformation was accompanied by a significant decline in poverty, especially in the high-growth economies of East Asia (Secular, 1999).

In many Asian countries, agricultural research played a key role in broad-based poverty alleviation by generating technologies that increased productivity, thereby improving the lives of tens of millions of farmers (ADB, 1997 and 1998). The "Green Revolution" that began in that period was a welcome remedy for the food shortages in Asia, Latin America and North Africa. Poverty rates in most East and Southeast Asian countries that benefited from the Green Revolution dropped significantly because of higher agricultural productivity and greater urban employment opportunities. In many cases, these changes resulted in higher living standards (ADB, 2001). Then agricultural growth fueled the demand for small-scale industry, which expanded off-farm employment. This in turn raised the demand for domestic agricultural production. This dynamic transformation process has been impeded in Africa by a lack of fundamental investment to get agriculture moving in the first place (Kodhek et al., 1999).

Although the success of development practices of countries such as Japan, Taiwan, Thailand, Malaysia and the Punjabis are largely perceived as examples of export-led growth, these countries gave agriculture a central place and deployed agriculture-based strategies. Japan and Taiwan can be cited as good examples in their economic progress from agriculture to industrialization. Boosting total factor productivity in agriculture by using improved seed-fertilizer combinations on small-scale farm units was their first step. For instance, between 1950 and 1966, impressive yield and labor productivity increases were made possible by means of full and efficient utilization of the agricultural labor force. This resulted in increased farm cash income, which in turn reinforced rural industrialization. Eleni et al. (1999) indicate that in Taiwan a 2.5% increase in the total factor productivity of crop production resulted in a

3.5% growth in crop production in the post-World War II period, and this helped transfer large resources from agriculture to other sectors. In effect, such resource transfers certainly meant a significant squeeze on agriculture, but even so, the rapid growth of output permitted a rise in per capita rural consumption of about 0.9% per year in this period.

The book edited by Mellor (1995) consists of a detailed analysis of eight countries or states. Six were highly successful in using agriculture as the leading edge of growth (Colombia, Costa Rica, Kenya, Punjab, Taiwan, and Thailand); one succeeded in agriculture but failed to obtain the multipliers for growth in non-agricultural income and employment (the Philippines), and one was a failure in both agriculture and non-agricultural growth (Argentina). Excluding Argentina, the seven countries saw economic growth of 6.4% on average, with agriculture, manufacturing and services growing by 4.5, 8.9 and 6.7% respectively. The per capita growth rate of agriculture determines the multiplier effects on other sectors and the seven countries averaged per capita growth in agriculture of 1.6%.

In his study, Mellor (1995) noted that there are many different routes to rapid agricultural growth. Thailand grew substantially through the expansion of land area with relatively low technology. In contrast, Punjab and Taiwan increased agricultural growth primarily through yield-increasing technology, as did the Philippines to a large extent. Colombia, Costa Rica, and Kenya obtained accelerated agricultural growth by changing the composition of production, and in particular from the rapid growth of agricultural exports. As a country's per capita income rose, the income-elastic demand for horticulture and livestock products allowed agricultural growth rates of 4% to 6%, based largely on domestic demand. Thus, rapidly growing domestic demand will eventually drive the shift in composition towards high-value agricultural products. Even in countries in which technological change was the dominant source of growth, changes in what was produced and increases in the amount exported played an important role.

In Taiwan, domestic demand played a far more important role than export growth in terms of labor absorption, particularly in the period from 1952 to 1986 (Park and Johnston, 1995). The rising rural consumption demand, which resulted from agricultural development, triggered the early development of Taiwan's small and medium enterprises, laying the foundation for a fast-learning, highly adaptive and dynamic sector. This continued to be of major importance as the country moved on to a later phase of development dominated by industrial exports (Lee, 1971; Eleni et al., 1999). As agricultural production and rural incomes grew rapidly in the 1960s, new small and medium enterprises also mushroomed rapidly in this period. Consequently, in the period from 1950 to 1985, Taiwan's economy was vastly diversified, leading to a fall in the relative share of agriculture from about 38% to nearly 6%. A decline in fertility resulting from the promotion of family planning, education and the reduction in infant and child mortality significantly helped Taiwan in the transformation process (Eleni et al., 1999). Taiwan's rate of growth of non-farm employment was consistently higher than that of the total labor force in this period, enabling the share of the agricultural labor force to decline from 56% to 15%. As a result, in 1974, employment in the non-agricultural sector outshone employment in the agricultural sector, thus accelerating the process of structural transformation (Moa, 1992).

Keeping rural people in rural areas was the other development strategy. In Taiwan, the proportion of the population remaining in rural areas was consistently high, in contrast to countries that were less successful in using agriculture to stimulate non-agricultural growth. Thailand was also notable in this respect, with a surprisingly high proportion of the population working in rural non-farm production jobs. Growth in agriculture has a stronger impact on the service sector, light manufacturing and construction than on what is more traditionally described as the manufacturing sector. However, in Punjab, non-agricultural growth was so dispersed that no dominant towns developed in the state. Spending resulting from rising farm incomes stimulates employment-intensive activities in small towns and rural areas, and it helps avoid the social and economic diseconomies of mega-cities (Mellor, 1995).

According to Kang and Ramachandran (1999), the most rapid growth of Korean agriculture was recorded from 1928 to 1937, with total growth in agricultural output being 41%. A combination of expanded irrigation, intensification of the use of chemical fertilizer, and modern seed varieties were the key factors driving growth, given the scarcity of arable land. Chemical fertilizer particularly was an important source of growth, contributing 42 to 77% of the overall growth in agricultural output. Labor and fixed capital was much less important. This ultimately created an agricultural revolution in Korea. Thus, the authors argue, the advancement in agriculture set the stage for economic expansion, thereby helping to create an industrial economy in Korea.

A paradox to today's development theory is the success of China in opposite periods of economic policy regimes. China achieved substantial and extensive institutional transformation both before and after policy reforms and was relatively successful in promoting growth and development. It experienced significant growth in total factor productivity, about 3.8%, and GNP growth of more than 8% per annum in both the 1952-1957 period of socialist central command policy and the 1978-1997 post-reform period of market oriented policy. The agricultural sector also responded very positively. According to Secular (1999), one feature both periods have in common is that investment was high, at 24% in the pre-reform period and 33 to 40% in the post-reform period. He therefore argued that investment was a sufficient condition in itself for rapid economic growth, regardless of the direction of institutional change 'within limits', and that the role of capital formation and institutional transformation depended very much on initial conditions. Hence, no single prescription will apply to all countries at all times. The recommendation by the IMF and World Bank, market liberalization for all at all times, can be best example of failure.

The Philippines provide an example of a major Green Revolution success, but one that was not enough to stimulate broad economic growth. Despite a rapidly growing agricultural sector, the Philippines experienced increasing rural poverty and declining real wage rates. In addition, while other countries experienced dispersion of the non-agricultural labor force throughout rural areas, the Philippines showed increasing concentration of the non-agricultural labor force in one city: metropolitan Manila (Mellor, 1995). Why the Philippines presented such a perverse record of multipliers from agricultural growth is a complex story. An important part of the story is poor macroeconomic policy that led to a concentration of investment in capital-intensive industries, particularly those located in metropolitan Manila. That pattern of development resulted in very little employment growth and starved the rural sector of the capital and institutional structures necessary to take advantage of growing rural incomes. Other factors including the structure of land tenure also played a role.

The case of Kenya shows that if agricultural growth is restricted to particular regions, the multipliers will be restricted as well. Kenya's experience also demonstrates clearly that the extraordinarily high population growth rate that characterized much of Africa can substantially undermine the effect that strong growth in agricultural has on the growth rate of other sectors (Mellor, 1995). The Argentinean case demonstrates how poor macroeconomic policy can interfere not only with agricultural multipliers, but also with agricultural growth itself. Still, sound macroeconomic policy alone is not enough either. The examples of Punjab and Taiwan show how critical institutional structures are to growth and what needs to be done to make them effective, not only for accelerated agricultural growth, but for strong multipliers as well.

Pyakurel (2002) argued that the focus of government policies on the development of infrastructure and input delivery alone also could not make any difference in agricultural development. For instance, despite its tremendous efforts in policy reforms and infrastructure development, Nepal failed to succeed in improving growth in the agricultural sector, mainly due to poor project design and implementation, institutional weakness, poor input delivery mechanisms and inefficiencies in the public sector agencies involved. Other reasons include lack of irrigation facilities, technology, land fragmentation and limited access to markets. Hence, Pyakurel (2002) stated, the mechanism of how to promote a market economy in Nepal is not clearly spelt out in the development strategy. He therefore suggested that policy-makers should review the plan thoroughly. In most cases, many sub-Saharan African countries share a similar experience of failures to achieve accelerated agricultural growth.

2.5.3 Agricultural Development in Sub-Saharan Africa: Failures and Progress

Two decades after the independence in the early 1960 were characterized by the high hopes for rapid African development, and much was of course achieved. However, by the end of 1970s, the major social and economic indicators began to decline. According to the Conference Advisory Committee (CAC) 2004, the drop in agricultural production resulted in massive imports (increased from 1.9 in 1970 to 6 billion US$ in 1980), while foreign debt had skyrocketed (from 5.4 in 1970 to 41.3 billion US$ in 1980). Consequently, the crisis in Africa was deepening; growth in agriculture was weak. The rate of growth in yield for major food grains was negligible between the 1960s and the 1980s. Labor productivity in agriculture became very low even by Asian standards, let alone by the standards of the developed world. The problems were caused by both external factors such as decline in terms of trade, oil crisis etc., and internal problems like political instability, inefficient resource use, environmental degradation etc.

Among many others, the response of World Bank and IMF to the Africa's problems is worth noting. It argues for the fundamental reform needed to stabilize the economy and reform the market to increase efficiency for investment and reinvigorate growth. As a result, SAP was introduced as of early 1980s as a move away from state-dominated development to reliance on the private sector, with the state retreating to setting appropriate policies. A significant number of countries have carried out successful programs of macroeconomic stabilization and structural reforms.

However, immediately following the devaluation of local currencies, farmers were faced with increased prices of purchased food and agricultural inputs, which disproportionately affected small farmers who had even less access to credit. Especially, the poor were adversely affected by the SAP program particularly in social services like health, education, employment etc. Moreover, farmers were constrained by poor infrastructure (especially markets and roads) and lack of technology. In addition, in this period there was persistent drought, which may also have partly obscured any positive policy effects (Mosley and Smith, 1989).

Two or three years after the reform of 1980s, aggregate supply price response analysis was carried out in eight African countries using time series regression by Bond (1983). Although the observation in that period showed that African farmers evidently responded to price incentives, analysis showed a statistically insignificant supply response to price in seven countries. Only Kenya was found to have a significant long-run supply response coefficient, but still only 0.16. Kenya had relatively higher agricultural potential and a relatively well-developed infrastructure that suggests such a capability.

Delgado et al. (1984) found that a major reason for the low supply elasticity of price in sub-Saharan Africa is the poor state of agricultural infrastructure and input distribution systems, reinforced by low input intensity, and the inadequacy of the existing technologies. Similarly, Hedhues, et al, (1992) underlined that the potential problem associated with supply is so immense; the non-price factors such as investment in education, health, transport, input supply, finance, research and extension constitute the binding constraints. CAC (2004) also argues that the major issue that remains in most countries is that the private sectors are weak to take up the role once performed by government marketing monopolies.

Only after several years of policy reforms could slight differences were observed in the structure of African economies (Kirk and Pillet, 1998). Although the great majority of sub-Saharan African countries are still in the early stages of structural transformation, there are significant differences among them. In terms of the size of the total and agricultural labor force and the total fertility rate, countries like Botswana, Ghana, Kenya and Zimbabwe are in a relatively better position. While Senegal achieved the highest degree of agricultural commercialization (29%), Uganda experienced the highest industrial growth (13%) in the 1990s, and Namibia achieved the highest agricultural value added per worker in the 1980s and 1990s.

In general, in the initial stage of the market reform, sub-Saharan Africa was beset with problems due to the fact that complex market arrangements were not following the reform. Moreover, the prevailing private and/or public product and financial markets had only a limited capacity to induce production or to reduce market risks for all economic agents, producers, consumer and traders (Beynon, 1988; and Beynon et al., 1992). Hence, the state of economic reforms in Africa appeared rather worse in the 1990s. Real GDP remained static while agricultural production fell substantially in per capita terms. Food production did better than non-food production, but even so, per capita food production was lower than in the pre-reform period. Unfortunately, agricultural production fared worst on average in the ten poorest countries, with four experiencing negative growth rates, while agriculture in Mozambique was in a state of virtual collapse. Delgado et al. (1984) suggested two categories of solutions to these problems: increasing producer price incentives through price liberalization and diminishing the size of the government's role and structural change (particularly improved research, input delivery, and infrastructure system), and emphasizing that one without the other is fruitless (Jayne, et al 2002).

The problem called on the need to a balanced approach between the state and private sector functions in development, still with emphasis on efficiency in resource allocation. As the result the Comprehensive Development Framework (CDF) and the Poverty Reduction Strategy Paper (PRSP) process were formulated in the late 1999 to integrate cultural, social, political and environment aspects into economic development strategies directed towards the pro-poor growth. The poverty emphasis implies that agriculture and rural development play a vitally important role in CDF/PRSP concept. The SAP and PRSPs have largely shaped regional approaches and country policies and their implementation in all of Africa (World Bank Group, 2001).

Despite the multifaceted problems, a promising scenario is that documented in the recent studies of Kodhek, et al. (1999). There are indications that there is an emerging trend of agricultural recovery and growth in some countries (e.g., Uganda). Sectoral policy reforms have begun to pay off. There has been an emerging trend of increased use of inputs and intensification in some countries. In such countries, agricultural productivity has grown. Nevertheless, the remaining problems are manifold. Even in cases where technologies have been developed, the ability of farmers to use them has been impeded by high costs of distribution, due to poor infrastructure, high transport costs, and policies that prevent private investment in the system.

On the other hand, the CAC remarks the crucial importance of peace and democratic leadership in the development process. According to the reviews by the advisory committee (CAC, 2004; and Jayne, 2002), most successful reforms have been in countries without war or civil unrest, where the reform process was characterized by strong political leadership combined with a commitment to reform under wide domestic participation and ownership. In countries that have been plagued by conflict and war (e.g., Ethiopia, Somalia, Sierra Leone, Rwanda, etc.), development has been pushed back by years, if not decade. The other key problem for many African countries is the unevenness of the playing field in international trade, with the European Union, the United States, and Japan protecting their markets against imports from developing countries, including Africa (Stiglitz, 2002). Hence, the New Partnership for Africa's Development (NEPAD), which was formed in June 2001 and set agriculture-led development strategy, emphasizes that the liberalization and globalization process needs to be fair and the playing field made more level.

2.6 Conclusion

Despite many arguments against agriculture, both developed countries and countries currently undergoing rapid growth have been benefited from the immense role of agriculture in the process of their economic growth and structural transformation. Failure to give due emphasis to agriculture chokes the food production system and undermines hunger alleviation as well as the transfer of resources necessary for the expansion of non-agricultural sectors, and ultimately slows or even prevents the process of structural transformation.

From the variety of approaches to economic development, one can glean certain factors that need to be considered in the economic development process: the focus on savings and investment and the critical importance of structural and institutional changes are not mutually exclusive. Their combined effect would enable strong sectoral linkages and, ultimately, changes in the structure of the economy to be achieved. Yet it is of paramount importance to consider the influence of the workings and structure of world economies in the world of economic globalization.

In order to promote efficient production and distribution through properly functioning pricing systems, the theory of neoclassical economics that advocates efficient pricing and market access is found to be an integral part of any successful development process, provided it is adapted to fit the unique social, institutional, and structural circumstances of the LDC concerned. Therefore, "getting prices right" is innately attractive as a quick and simple remedy for the poor performance of agriculture. In addition, the structural paradigm that advocates technological innovation in agricultural production is the other, more important approach to solving the supply problem. Central to this process are developed agricultural research and extension systems, improved rural infrastructure, transport, and input and output markets, and the development of a rural financial sector to facilitate resource mobilization. If these conditions are fulfilled, agriculture has the capacity to exploit productivity-increasing technological innovations that make large net additions to national income and hence to purchasing power. Each of the successful cases discussed in the literature shows that modern science would allow agricultural growth to accelerate more quickly.

While recognizing that industry is important for pulling Africa out of poverty, only industrialization based on strong agricultural growth can lead to broad-based poverty alleviation of the kind seen in parts of South-East Asia. A comparison of rapidly-growing Asian countries with sub-Saharan African countries showed that increasing agricultural income, whether based on expansion of resources or productivity-enhancing technologies, is the single most important factor in explaining differences in living standards, enterprise expansion, and job creation in the non-farm sector and overall economic growth. Thus, fostering agricultural productivity, raising the share of market surplus rather than subsistence is also critical for Ethiopia in all its efforts to achieve agricultural and then overall economic structural transformation.

3 Review of Ethiopian Economic Development Policies, the Position of Agriculture in the Economy and the Challenges

Government policies and development strategies have a significant impact on the economic growth and development opportunities of a nation. Sound and prudent policies and strategies serve as a vehicle for development, whereas ill-conceived and inflexible government policies and strategies hinder development and may eventually even lead to poverty. Ethiopia is one of the poorest countries in the world: poverty in the country is among the worst anywhere. The miserable economic condition of the country is reflected in every sector in all social and economic indicators.

Ethiopia has experienced three distinct government regimes with completely different economic development policies and strategies (Box 3.1). These are the feudal-bourgeois regime, which pursued a largely pro-capitalist and pro-feudal economic policy (before 1974), the centrally planned socialist economic regime (1974 to 1991), and the pro-capitalist regime under the Ethiopian People's Revolutionary Democratic Front (EPRDF), (1992 to the present). The objective of replacing the feudal-bourgeois system by the revolutionary socialist system was to reorient the Ethiopian Economy along socialist lines in the hope of releasing the productive forces of the country to bring about more rapid economic growth on more egalitarian grounds, significantly transforming the structure of the national economy.

From 1992 onwards, the EPRDF government replaced the centrally planned socialist system with a market-oriented economic system. The government has adopted the SAP advocated by the World Bank and the IMF. Consequently, various policy reforms have been undertaken to achieve the program's immediate objective of macroeconomic stability (stabilization policy), and the long-term objective of sustainable economic growth. In addition, sector-specific and broad-based economic policies and development strategies have been implemented to complement and support the macro policies. Agricultural Development-Led Industrialization, or ADLI, is the principal guiding strategy of the government.

More than a decade has now passed since Ethiopia embarked on a new course of economic policy reforms towards what can best be described as a new liberal and market-friendly regime. This chapter briefly reviews major development policies implemented by earlier government regimes, with the principal focus on the development strategies carried out by the current government, the EPRDF. It also describes the performance of the aggregate economy, and takes a closer look at the agricultural sector in particular.

Box 3.1: History of Ethiopian Development Strategies

1950s
System: pro-capitalist and pro-feudal (feudal-bourgeois).
Development strategy: export promotion (outward-oriented).
Objective: to earn high levels of foreign exchange through diversified exports and trading partners across the world. Higher export revenues were to be achieved through better quality and larger quantities of exports in order to finance imports.
Measures implemented: incentives to attract foreign investment.
Results: unsuccessful; policy was biased against agriculture in favor of industry and cash crops and resulted in food shortage and famine.

1960s
System: continuation of pro-capitalist and pro-feudal (feudal-bourgeois) system.
Development strategy: import substitution (inward-oriented).
Objectives: to save foreign exchange by producing goods domestically that would otherwise be imported.
Measures implemented: domestic industries were protected from outside competition by imposing high import tariffs and taxes, restrictions on types of goods and quotas, based on the "infant industry argument".
Results: Unsuccessful because of the heavy dependency of already existing industries on imported raw materials and spare parts. Domestic industries remained at infant stage until the successor political regime introduced the socialist system in 1975.

1970s and 1980s
System: socialist (command economic system).
Development strategy: import substitution was pursued.
Development principle: rapid economic growth based on egalitarian principles.
Measures implemented: confiscation of private property, i.e. extra houses and large firms operating in the country; nationalization of rural and urban land; land placed in state ownership. Rural land then distributed to each tiller. Restriction of private enterprises and promotion of state enterprises, associations, cooperatives and organizations based on the socialist system by providing subsidies on both inputs and outputs. Markets were predominantly controlled by parastatal organizations such as the Agricultural Marketing Corporation (AMC) and the Ethiopian Domestic Distribution Corporation (EDDC).
Results: the economy totally collapsed.

1990
Development strategy: a mixed economy was introduced.
Principle: allowing private sector participation within a socialist development policy framework.
Result: strategy was not implemented because of the severe civil war between the socialist government and the then rebel forces, who took over the reins of government in mid-1991.

1992
System: pro-capitalist.
Development strategy: rural-based and export-oriented.
Principle: private ownership of productive resources and free market-based development; however, factor markets not fully liberalized.
Measures implemented: introduction of Structural Adjustment Policies (SAP) and promotion of privatization, reducing the role of the state in the production sectors. New development strategy formulated, referred to as ADLI, or Agricultural Development-Led Industrialization.
Results: under investigation…

Source: Collected from different reports on Ethiopian development studies and personal observation.

3.1 Outline of Economic Development Policies and Strategies and Problems Encountered

3.1.1 Pre-Reform Period (before 1991/92)

To trace the history of Ethiopian economic development strategies, one needs to outline the situation since the conclusion of the World War II (Box 3.1). Here, the discussion is confined to the salient features of the development strategies followed. Until the 1960s, the economy was by and large open to the rest of the world: the government pursued an export-led development strategy. The major components of exports were hides, skins, and coffee. Consequently, an increasingly large area of agricultural land was brought under cultivation to produce crops for export, and hence much of the physical infrastructure, such as expansion of the road network, was linked to the growth of foreign trade. The classical goal of trade was to mobilize government revenue by imposing taxes on exports and imports. The same trade policy was maintained under the socialist regime, despite the ideological changes, and the principles of protectionism and a fixed official foreign exchange rate were left intact. In general, restrictive exchange rate controls and a cumbersome import tariff regime limited the nature, volume and value of exports and imports in the pre-reform period. In addition, the high level of resources committed for the development of exports of goods and services resulted in significant neglect of the promotion of food products. This in turn brought about a serious countrywide problem of food shortage and famine.

Subsequently, from the early 1970s to the late 1980s, the inward-looking (import substitution) strategy was pursued particularly with a view to creating favorable conditions for domestic industrial development and effectively protecting it from external competition. The country rigorously implemented a development strategy of import substitution by introducing high import tariff rates and directly restricting quotas, and/or banning imports of certain commodities in general, with the intention of expanding and encouraging domestic industries. During the bourgeois period, there were tax incentives, and credit was provided on favorable terms, particularly to medium and large industries. Foreign and domestic investors were freely involved. Small-scale industries, however, were neglected. Poor infrastructure facilities and lack of an articulated economic policy were the major problems that limited the expansion and contribution of the sector (MEDaC, 1999). Under the socialist system too, the government followed the import substitution strategy of the bourgeois regime. In its allocation of resources, the government emphasized only the development of the large and medium-scale industries. Only later, in its Ten-Year Perspective Plan (1984-1994), did the government recognize the importance of promoting small-scale (light) industries, and established the Handicrafts and Small-Scale Industries Development Agency. Nevertheless, the agency failed to bring about development of the small-scale industries, mainly due to institutional weakness (MEDaC, 1999). However, apart from such general policy guidelines, there was no definite development strategy for the industrial sector in the pre-reform period.

Likewise, very limited consideration was given in this period to the development of the agricultural sector. Different agricultural development projects were implemented one after the other, and sometimes simultaneously. Three important projects were well documented: the "Comprehensive Package Projects" (CPP), the "Minimum Package Projects" (MPP), and the Peasant Agricultural Development Extension Project (PADEP). Despite their differences in approach, all these projects had one common objective, namely to increase farm productivity and farmers' income.

Comprehensive Package Projects (CPP): The first major development efforts to raise agricultural production and productivity were initiated in the late 1960s and early 1970s during the feudal-bourgeois period in selected areas of high agricultural potential, notably Chilalo, Walayta and Adaá. The three CPP were named according to their location: Chilalo Agricultural Development Unit (CADU, later renamed ARDU), Walayta Agricultural Development Unit (WADU) and Adaá District Development project (ADDP). These projects were launched in Ethiopia following the "Green Revolution" of Mexico and India. The projects were aimed at integrating agronomic research, disseminating improved seeds, chemical fertilizers, and improved farm tools, providing marketing and credit facilities, and promoting rural road construction and the supply of potable water. The basic intention was to improve the living standard of all farmers without excluding the poor peasants, and these packages were able to demonstrate the potential of increasing yield. However, only the then landlords, businessmen, and some government officials in a position to set up commercial farms were able to adopt the package recommendations. Small and poor farmers, on the other hand, found the recommendations too costly to adopt. Ultimately, the project ended up creating rural elites (exacerbating income disparity). Moreover, the few adopters (beneficiaries) mechanized their farms and dramatically increased and expanded commercial farming, which resulted in the eviction of small farmers (Dejene, 1990).

Minimum Package Projects (MPP): three years after CPP, a Minimum Package Project was designed. The objectives of the project were to be achieved through diffusion of a few proven improved methods of agricultural production. The attempt was to achieve wider coverage with the diffusion of selected farm implements, fertilizer and improved seeds developed by the CPP. The project was undertaken in two phases, Phase I in 1971-19801 and Phase II in 1981-1984, and thus stretched from the feudal-bourgeois period to the socialist period. In the first phase, like the CPP, the project was confronted with the problem of land tenure, which prevented poor peasants from benefiting (Cohen, 1974).

Following the overthrow of the imperial bourgeois system by Socialists in 1974, government policy concerning agriculture changed in line with the ideological thinking of socialist principles. The new policy was characterized by complete state ownership of land, and nationalization of all private and commercial farms. The large commercial farms were recognized as state farms, and all rural land was brought under public ownership. Land was freely redistributed to small farmers, who were granted usufruct rights to cultivate it. Private investment in the agricultural sector was prohibited, whereas collectivization of peasants into producer and a service cooperative was greatly encouraged. Another important feature was that the government controlled all agricultural markets, setting obligatory food grain quotas, predetermining low prices for deliveries, and restricting the movement of agricultural outputs from region to region within the country; such aspects were under the control of a parastatal marketing firm known as the Agricultural Marketing Corporation (AMC).

Collectivization of farmers into producer cooperatives was thought to be the solution to the problem of diminishing farm size, land fragmentation and tenure insecurity that might arise as a consequence of the land reform policy. The government then offered the cooperatives special incentives, such as cut-price for fertilizer, a higher price for grains, and lower land tax and interest-free loans (Mulat, 1999). In the mid-1980s, a villagization program was launched, as this was considered the primary condition for successful collectivization. Thus, the second phase of MPP concentrated more on the cooperativization and collectivization process than on bringing improved technologies to the individual farmers. Ultimately, the project was also handicapped by insufficient manpower and financial resources, and hence failed to achieve its objectives.

Peasant Agricultural Development Extension Project (PADEP): PADEP, designed to reflect socialist principles, was launched in the mid 1980s following the failure of CPP and MPP. The primary objectives of the project were to increase food production and promote service and producer cooperatives. However, this project was not fully implemented because of financial constraints and because of the unnecessarily heavy involvement of the government in the grain market, which the potential donors could not accept (Mulat, 1999).

As documented in the works of Tennase (1985) and Assefa (1995), all of the above projects failed to achieve their aims because of both external and internal problems. The main external problems included inappropriate project design and poor research-extension linkages, while some of the internal problems were very low development agent-farmer ratios, shortage of qualified experts and inadequate financial resources. Generally speaking, the efforts of the centrally planned socialist system, with their ill-conceived interventions, merely succeeded in stifling private initiatives for planning and implementing development programs and, as a result, development of the agricultural sector failed to materialize. Instead, the socialist policies tended to create bottlenecks to improving the productivity of the small-scale subsistence farmers that constitute the bulk of Ethiopia's agricultural production system.

3.1.2 Post-Reform Period: Outward-oriented and Rural-based (1991/92 onwards)

Since the seizure of the power by the EPRDF government, radical policy reforms have been implemented, reversing the socialist system and replacing it with a liberal market development strategy (outward-oriented) rooted in rural agricultural development. This was aimed at revitalizing the economy and stimulating growth from the low level of development resulting from the inappropriate and inflexible economic policies of the socialist era. Far-reaching policy reforms and sector development strategies have been formulated in order to rectify the policy constraints of the earlier system.

3.1.2.1 Major Macroeconomic Policy Reforms Undertaken

Fundamental Economic reform programs, Stabilization and Structural Adjustment Programs have been undertaken in order to remove the structural bottlenecks in the economy. The major reform measures include exchange rate devaluation, market liberalization and price deregulation, privatization, transport deregulation, elimination of subsidies, and reduction of tax rates.

Exchange rate devaluation: - to promote the export sector in particular and to encourage efficient allocation of resources, one of the first steps of the economic reform was devaluation of the exchange rate by 58 % in US$ terms from 2.07 birr/US$ to 5.00 birr/US$ in 1992. The government also introduced a bi-weekly foreign exchange auction market from 1993. The local currency devaluation continued, reaching an average of 8.65 birr/US$ in mid-2004.

Fiscal policy reform: - suspension of taxes and duties on all export goods (except coffee) is another related reform, intended to provide incentives for exporters and stimulate production of exportable goods. In the context of trade liberalization, the government reduced the level and dispersion of effective rates of protection, average nominal tariff rates. The shift from pursuing a more or less protectionist regime, inherited from the Socialist regime, to a more liberalized policy drastically reduced custom duties from an average of 230% to an average of 50% of the value of an item and all export duties were eliminated. In order to improve the budget balance, the government has totally removed subsidies, and increased the tax base (source of tax income) by introducing other types of tax, such as tax on rental and interest income, and value-added tax, and the process is subject to regular review.

Trade (market) liberalization and price deregulation: - the government lifted restrictions on trade by introducing a system of open general licenses. To encourage producers and consumers to allocate their resources efficiently in a free market system, price controls on goods and services produced by private and public enterprises and agricultural products were abolished, and market mechanisms were allowed to determine them. The AMC has been removed, and the output and input market have been liberalized to allow them to operate freely and to encourage competition, efficiency and profit maximization. Zonal restriction of transport services has been also lifted.

Even if the government opens up the agricultural marketing, the fertilizer market in particular has not yet been totally liberalized. Directly or indirectly, the government has maintained a trade monopoly over fertilizer import and distribution by enterprises established by its political parties. These include Ambasel, Dinsho, Wando and Guna; they have been dominated the import and retail of fertilizer.The seed enterprises have also been similarly owned only by the government. It might be from distrusting the private sectors as fertilizer is strategic input in the developments strategy or simply for political manipulations. This may discourage privates sectors from investing in import and distribution of such modern inputs holding back the full-scale privatization and liberalization and hence the efficiency of agricultural marketing. For instance, the collapse of the private fertilizer agency, known as Ethiopian Amalgamated-Limited, is a typical incidence that resulted from unfair treatment by those government affiliated fertilizer enterprises. In addition, because markets are still underdeveloped and access to market information is poor, it remains questionable whether the markets are in a position to guide efficient allocation of resources.

Monetary policy: - with the objective of stabilizing the macroeconomic environment and creating an effective and efficient financial sector, the government has progressively introduced indirect monetary instruments. In addition to the removal of subsidies, banks adjust interest rates according to the changes in the economy, and open market operation has been introduced for government securities (treasury bills), maintaining positive real interest rates for both depositors and lenders. The other important reform is the introduction of a competitive financial sector, which includes the establishment of private banking and private insurance companies, and the establishment of micro financing institutions specialized in providing credit to emerging small and micro enterprises and farmers. However, foreign banks have not yet been involved in the financial institutions of the country.

Privatization: - in order to enhance productivity and profitability, many public and state-owned enterprises have been privatized, and the process of privatization is still under way. The government started by completely halting development activities relating to collectivization and villagization in the rural, and public corporations has been drastically reduced in the urban. Instead, smallholders and private commercial farms have been encouraged in the agricultural sector. Furthermore, the role of state has been temporarily limited to large and medium-scale industries such as pharmaceutical firms, metallurgical and power plants, which the government assumes the private sector to be unable to manage in the short run. Some public enterprises that have been retained in state and/or public ownership have been given managerial autonomy. In the agricultural sector, the new government decided to hand over some of the state farms to nearby farmers and to privatize some of them. Up to now, only 13 out of 26 state farms remain under state control, producing mainly wheat, maize, cotton, coffee and tea on an area of only 156,040 hectares.

In addition, investment law has been improved so as to encourage both insiders and outsiders to freely invest in the economy with no maximum capital limit. This is primarily aimed at expanding and diversifying production for domestic use as well as for export and thereby to create employment opportunities. Moreover, it is expected that the country might benefit from incoming foreign technologies and know-how. However, it is obvious that investment (especially direct foreign investment) chiefly depends on the political and economical stability of a country, which in Ethiopia is still lacking at the present time.

Land tenure: - in spite of all the reforms mentioned above, land has remained under state control. According to the 1994 constitution of the Federal Democratic Republic of Ethiopia, the right to own rural and urban land and any natural resources is vested exclusively in the state and the people of Ethiopia. The constitution clearly states that land is the "common property of the nations, nationalities and peoples of Ethiopia" and shall not be subjected to sale or to other means of transfer. Nevertheless, this land tenure policy is the subject of great controversy at present, and many economists and other professionals argue against it based on the negative impact it may be expected to have on efficient use of farm resources and on promotion of modern, large-scale production systems in agriculture. Furthermore, state ownership of land is strongly criticized because it discourages any land-improving investment and can induce over-utilization (degradation) of natural resources.

3.1.2.2 The New Economic Development Strategy of the Current Government: Agricultural Development-Led Industrialization

The EPRDF government has three interrelated objectives: sustainable economic growth, interregional equity, and self-reliance, in other words, dependence on national resources and independent national development (MPED, 1993). To this end, the government has formulated a new development strategy known as the ADLI. ADLI bases its root on agriculture as this sector is the mainstay of the economy; around 85% of the population are engaged in it, and the livelihoods of the majority of the poor are concentrated in this sector. In general, the government has accepted agriculture as an engine of growth on account of its potentially superior growth linkages, and its potential for generating surplus, creating markets, and providing raw materials and foreign exchange for the industrial sector.

As its name implies, ADLI's stated aims are to promote agricultural transformation as the first step, and subsequent industrialization (economic structural transformation) in the long run. While structural transformation can be brought about gradually, the primary objective is to attain high productivity growth in the agricultural sector in order to tackle the problem of food shortage in the country by improving the productivity of smallholdings and large and small private commercial farms. Such a process is expected to raise the income of the poor and enhances export earnings. Ultimately, it is anticipated that growth in the agricultural sector will create and strengthen sectoral linkages in the economy. It is expected to be a dynamic and self-propelling process that progresses under its own momentum, adapting to create forward and backward linkages between agriculture and industry. To be self-reliant, the agricultural sector should supply raw materials for the manufacturing sector and create market outlets (effective demand) for the outputs of the manufacturing and other sectors of the economy by increasing smallholder incomes. As the living standard of the rural population improves, the need for domestically manufactured goods, farm inputs and other secondary and tertiary products would be likely to rise, and this could induce industrial expansion.

At program level, ADLI embraces an export-led development strategy and incorporates parallel and coordinated development of agriculture and industry. It consists of a number of components encompassing almost all sectors of the economy. The main components of ADLI include primarily agricultural development, promotion of industry, education, and development of infrastructure (road construction and development of communication networks). However, ADLI requires that both forward and backward linkages be sufficient to generate the needed multiplier effects and that both agriculture and industrial supply are sufficiently responsive or elastic in response to changes in demand. However, analysis made by Adelman and Vogel (1992) using the Social Analysis Matrix for Sub-Saharan Africa showed that agriculture's backward linkages are considerably greater than its forward linkages. The supply responsiveness of the non-agricultural output to the growth in agriculture can only satisfy a fraction (0.16) of the induced demand for manufacture output. This is virtually so weak and when added to the poor performance of agriculture, the applicability of the ADLI strategy depends on the ability to initially address these supply-side considerations, particularly the capacity to increase the productivity and agricultural production.

3.1.2.3 Development Strategy for the Agricultural Sector

Smallholder farmers are the main source of agricultural production for both domestic consumption and export products. Food security and a higher income level for this section of society and for the nation in general could be achieved by raising farmers' resource productivity. This can be attained mainly by introducing new and improved technologies, constructing small-scale irrigation schemes, developing roads and supporting indigenous and foreign investment in agricultural production. In order to promote rapid growth in agriculture, therefore, a new system of agricultural extension referred PADETES was designed in 1994/95 within the framework of ADLI. The extension system combines the extension management principles of the Training and Visits (T&V) system with the Sasakawa-Golbal 2000 (SG 2000) extension system. Since its inception, it has been implemented on a massive scale, covering almost all agro-ecological areas in virtually all parts of the country. Following the launch of the program, extension activities were undertaken intensively until 2001, but then entered the doldrums because of the sudden collapse in grain prices in 2001 and the incidence of drought in consecutive years that followed. Ultimately, these circumstances created resistance on the part of farmers to participate in the extension activities.

The main elements of the extension package were fertilizer, improved seeds, pesticides and better cultivation practices, principally for cereal crops (maize, wheat, tef, barely, sorghum and millet) (Mulat, 1999). Under this extension system, the district bureaus of agriculture provide participating farmers with a package of inputs. Participating farmers allocate ¼ to ½ a hectare of their land for a demonstration plot and pay a 25% down payment on the input package at the time of planting, with the balance due after the harvest. The farmers managed the demonstration plots under the supervision of agricultural extension agents.

It is worth mentioning here that Mulat (1999) notes a common feature shared by all the previous and current agricultural development strategies. He underlines that all agricultural development strategies and projects implemented under the imperial and socialist system, and under the current government, were heavily dependent on external donors for funding and technical assistance. In the new extension program, for instance, the majority (78%) of the finance in 1995/96 and in 1997/98 for importing chemical fertilizer, the most important input promoted by the extension program, came from IDA credit and grants. It appears that the extension service can operate only if donors are willing to put up the necessary foreign exchange for importing fertilizer (Mulat, 1999). This implies that the government lacks self-reliance in its endeavor to enhance agricultural development. In addition, Mulat emphasizes that the local and national private institutions necessary to provide research, extension and input marketing remain underdeveloped

In order to complement and accelerate the development of agriculture, reforms in the policy and development strategies for industry may be indispensable. Nevertheless, aside from the fundamental public enterprise reform, there is no clearly formulated development strategy for the industrial sector. In the development strategy of the government, only a series of suggestions have been provided, and no articulated strategies are indicated. In line with the general economic development strategy of ADLI, the suggestions provided relate to resource use intensity and the composition of outputs to be produced. It encourages the use of labor-intensive technologies and extensive use of domestic raw materials to absorb surplus labor from the rural and urban sectors and to create effective demand for the products of the agricultural sector. Owners of manufacturing plants are motivated to produce outputs for which there is a large effective demand in the domestic market, i.e., outputs should meet the needs of the residents and consider the purchasing power of the people. These suggestions imply the expansion of cottage and small-scale industries as a key prerequisite strategy for agriculture and industrial linkages that would provide an impetus for structural transformation of the economy. In addition, the promotion of industries that rely largely on local raw materials, such as the sheet glass industry, and cement and chemical factories, is encouraged. Industries that depend on imported raw materials are encouraged to produce capital (intermediate) goods.

Improving rural education and transport infrastructure are the other main areas of ADLI. Emphasis is placed on expanding the education system at all levels, with particular attention to primary schools in conjunction with services provided to the rural population. Maintaining, upgrading and building new high roads and feeder roads in rural areas have been given well-deserved attention in the development strategy.

3.2 Government Expenditure on Agricultural Development Compared with the Industrial Sector

Total sector development expenditure, the sum of capital and current expenditure, and expenditure allotted to a particular sector or sub-sector were considered in this particular study. The sum allocated was also a useful point for consideration, as the government sometimes reshuffles one with the other. In the post-reform period, special emphasis has been given to the development of the social sector, infrastructure and the agricultural sector. Although it is not possible to ascertain the exact amount of money spent on each sector, financial assistance from bilateral and multilateral organizations has constituted an important share, supplementing the government budget for development expenditure.

As stipulated in the policy reforms, the government has cut spending on industry and other sectors like trade and tourism in the expectation of wider involvement of the private sector, so that substantial resources have been released for spending on social, infrastructural and agricultural development. Hence, spending on agriculture, road construction and education has increased sharply while spending on the industrial sector has fallen over the post-reform years (Figure 3.1 and 3.2).

On average, Government expenditure on agriculture and road construction increased from about 2.2% and 1.0% of GDP respectively in the pre-reform period (1975-1991) to about 3.1% and 1.6% of GDP in the post-reform period (1993-2000). No change was observed in the proportion of expenditure on education, which remained at 3% of GDP in the two periods mentioned.

Comparing expenditures in 1992 and 2000, in absolute terms, these amount to an increase of about three times in agriculture and education, and over ten times in road construction, but a fall of three times in industry. This implies that the government completely trusted the private sector to fully participate in and promote the industrial sector immediately after the reform. On the other hand, however, this fall in government development expenditure may partly reflect the government's underestimation of the contribution of the sector, and its overestimation of the agricultural sector in the economy as a whole. Given that the process of privatization is in its infancy, and the legal system pertaining to property rights is not strong, as long as the two sectors are interlinked and for the reason that the ultimate goal of development is industrialization, such an imbalance in the country's development strategy may be counterproductive.

Spending on agriculture was committed mainly to agricultural research, extension of improved technologies and staff training for front-line development agents. The spectacular change in spending on road construction was used for rural road construction and major road maintenance and rehabilitation (MEDaC, 2000). Apart from wages and salaries, education expenditure was used particularly for expansion of elementary schools in rural areas. Consequently, enrolment in primary schools in 2000 rose to three times the level recorded in the early 1990s. Despite considerable efforts that were also made to expand colleges and universities in different regions of the country, student-teacher and student-school ratios deteriorated continuously (MEDaC, 2003). For example in primary schools (grades 1-8), the ratio was about 33:1 in 1995/96, falling drastically to about 60:1 in 2002. The concentration of students in a school was also radically increased.

Figure 3 1: Annual Total Government Expenditure on Sector Development (millions of birr)

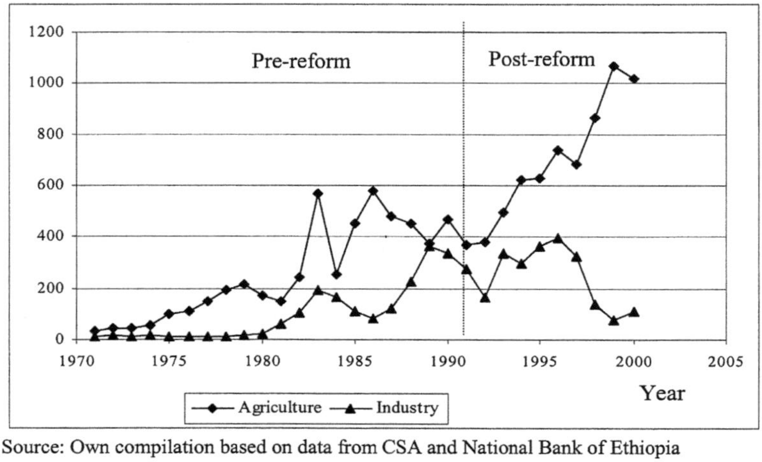

Source: Own compilation based on data from CSA and National Bank of Ethiopia

The student-school ratio in primary schools rose from 344:1 in 1995/96 to 674:1 in 2001/02 and in high schools from 1164:1 to 1681:1 in the same period. Such disproportionate growth in student and teacher/school ratios would inevitably reduce the quality of education unless compensated for by improvements in facilities such as books and other teaching aids. Unfortunately, the latter is the biggest constraint faced by the Ministry of Education in general. Although the impacts in the longer term are vivid, development of education and road networks in rural areas is expected to enhance resource transfer and productivity, and product and input marketability in the production sectors, primarily the agricultural sector.

Figure 3.2: Annual Government Expenditure on Education and Road Construction (millions of birr)

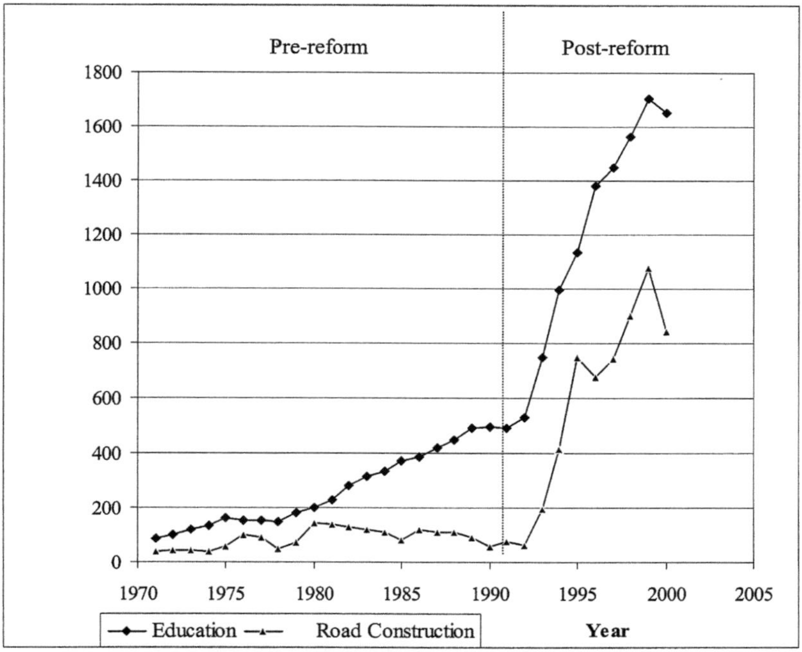

Source: Own compilation based on data from CSA and National Bank of Ethiopia

3.3 Assessment of the Performance and Contribution of Agriculture to the Economy in the Post-Reform Period

3.3.1 Growth and Relative Contribution of Agriculture to the Economy as a Whole[4]

One might expect that the policy reforms would result in significant growth in each sector, which in turn should ultimately lead to changes in the structure of the economy. In Ethiopia, however, no such striking change was observed. Table 3.1 summarizes the growth rate and the relative contributions of agriculture and the other sectors to GDP growth. Considering only the period from 1975 to 2001, real GDP estimated at constant factor cost (1980/81) was growing at about 2.78% per annum, with an appreciable difference between the pre- and post-reform periods. In the post-reform period, it grew at a rate of about 4.23%, which is more than double the growth rate experienced in the pre-reform period (2.05%). Compared to 1990, GDP increased by 53% in 2001. This outcome was also reflected in the growth rate of real per capita GDP. The declining trend in per capita GDP in the pre-reform period was reversed to give a modest positive growth rate of 1.34% in the post-reform years.

Because of the severe drought in 2002 and 2003, however, GDP fell by 1.6 and 11.4% respectively, while per capita GDP slumped by 3.8% in 2003. Moreover, agricultural GDP fell by about 3 and 12% respectively in 2002 and 2003, with the brunt of the fall being in food production and agricultural growth. Thus, the actual increment in the ten-year period from 1993 to 2003 was actually unimportant from the policy reform perspective. Taking the period from 1975 to 2003 as a whole, no growth in per capita GDP can be discerned (see Figure A3.1 in the Annex).

Agriculture maintained its supremacy in the economy at least until recently (2001), when its share became almost equivalent to the share of the service sector. Towards the end of the socialist epoch, in 1990, the share of agriculture was about 50%. The share of the agricultural sector fell at a rate of 2.7%, reaching its lowest point of 39% in 2003. Overtaking the agricultural sector in terms of relative importance, the share of the service sector was growing even faster (at a rate of about 3%), accounting for 49% of the GDP in the same year. Nevertheless, the share of the industrial sector unfortunately remained stagnant at about 10 to 11% throughout the whole period. Table 3.1 summarizes the details of these developments.

[4] In assessing the performance of the sectors and the economy as a whole before and after the reform, information relating to 1991/92 was purposely excluded, as the data would not represent a normal situation due to the severe civil war between the then rebels (the present government) and the socialist regime. Data from 1990 were used as the basis for comparison of the performance of selected economic indicators after the reform.

The growth in the GDP was further disaggregated into sectors to understand the extent to which the performance of each sector contributed to overall economic growth. Considering the effective period of the reform (1993 to 2003), analysis shows that, except in the service sector, the performance of both the agricultural and the industrial sectors was restricted to trivial progress. Consequently, only little change prevailed in the structure of the economy after reform. Growth in GDP is the weighted average of growth in the agricultural, industrial and service sectors. Using a mathematical model (Chiang, 1984), this can be shown using the following identity:

$$rY_t = \alpha_a rY_a + \alpha_i rY_i + \alpha_s rY_s$$

where, α_a, α_i and α_s are the weighted share of agriculture, industry and services sectors respectively in the specified period; and rY_a, rY_i, rY_s, and rY_t are the compound growth rate[5] of agriculture, industry, services and GDP, respectively. Terms on the right hand side constitute the growth contributions of each sector to overall growth in GDP. The percentage share of a sector's growth contribution to the overall economy can also be easily computed from the identity. For example, the percentage growth contribution of agriculture to the overall economy is computed as $(\alpha_a rY_a/rY) \times 100$.

Agriculture was growing steadily, with annual ups and downs owing to climatic conditions. In fact, the performance of agriculture after the reform was relatively better than in the pre-reform period. While it was almost stagnant (0.35%) prior to reform, excluding the drought year 2002/03 it grew at a rate of 1.55% after the reform (significantly different at the 10% level). Accordingly, agriculture contributed on average about 17% to the growth of GDP in the post-reform period, compared to only 9% in the pre-reform years. So, although agriculture constitutes a significant share of the economy and much more could be expected, its performance and contribution to the overall economy after the reform was satisfactory. Because agriculture in Ethiopia is characterized by predominantly traditional methods of farming, with farming practices having changed little over the past decades, its output is very susceptible to natural changes such as rainfall.[6] Almayehu (2001) also pointed out that the growth of agriculture and GDP move together rhythmically, indicating that the economy is heavily dependent on rain-fed agriculture.

5 A compound growth rate is defined as: $y_t = y_o(1+r)t$ where y_t and y_o are output in the time t and r is the annual compound growth rate. It can be estimated using the least square method and applying a log-linear model: $Lny_t = \alpha + \beta t$ where $\alpha = Lny_o$ and $\beta = 1+r$. Then the exponent of estimated β-1 gives us the compound annual growth rate r.

6 In 2000/01, agriculture recorded the highest growth (11.5%) and contributed about 65% to real GDP growth; real GDP also grew at 7.7%, mainly due to a bumper harvest, which in turn was the result of good weather conditions, and more intensive use of chemical fertilizer and improved seeds.

The growth performance of the industrial sector in the post-reform period was not significantly different from the earlier period; it was growing at an approximately similar rate. Surprisingly, its relative contribution to GDP growth was dropped by half, from 25% to 12%. The most remarkable changes were noted in the service sector. In the post-reform period, this sector was growing at a rate of 7.23% per annum (significantly different at 5% level), which is about double the growth rate it experienced prior to the reforms (4.14%). Hence, it contributed enormously to the growth of GDP, around 71%, in the post-reform period. In general, the service sector's contribution outstripped that of the agricultural and the industrial sector in terms of both absolute and relative growth rate of the real GDP.

To briefly understand a growth linkage of the sectors, a simple test of causal relationship was undertaken using a *Pair-wise Granger Causality* test technique. The result shows that the growth in the service sector has been found to granger cause the growths in the agricultural and industrial sectors. On the other hand, there has been no significant causal relationship existed between the growth in the agricultural and industrial sectors, and neither these two sectors have influenced the growth in the service sector (see Table A3.1 in the Annex). Particularly, absence of the causal relationship between agriculture and industry seems to contradict to the theory of economic development. It suggests further research and policy attention.

Table 3.1: Sectoral Contributions to GDP Growth and their Relative Importance (at 1980 constant factor cost)

Economic Indicators	1990	1993	1995	1997	1999	2001	2003
Real GDP per capita (birr)	242.6	225.2	231,4	251,9	248,0	265,6	244,6
Real GDP (millions of birr)	11350	11799	12645	14640	15294	17354	16904
Agriculture (%)	51	54	50	51	45	45	39
Industry (%)	11	10	11	10	11	11	12
Services (%)	38	36	39	39	44	44	49
Average Growth Rate (%)*	**1975-1990**		**1993-2001**		**1975-2003**		
Real GDP per capita	-0.28		1.34		0.00		
Real GDP	2.05		4.23		2.78		
Average share of agriculture	55.6		47.1		52.4		
Absolute growth in agriculture	0.35		1.55		1.51		
Contribution to GDP growth (%)	*9*		*17*		*28*		
Average share of industry	11.6		10.9		11.1		
Absolute growth in industry	4.54		4.81		2.89+		
Contribution to GDP growth (%)	*25*		*12*		*12*		
Average share of services	32.9		42.0		36.4		
Absolute growth in services	4.14		7.23		4.54		
Contribution to GDP growth (%)	*66*		*71*		*59*		

Source: Computed from reports of the Ethiopian National Bank Quarterly Bulletin, various issues.
Note: * Log Linear method is used; + Because of the inclusion of the years 1991 and 1992, the overall growth rate of the industrial sector is less than the average of the two periods.

3.3.2 Changes in the Performance of Food Grain Supply after Economic Reform

In Ethiopia, almost all food for the population comes directly from the agricultural sector. According to CSA (2000) and MEDaC (1999), crop production alone contributes over 90% of the domestic food supply, of which smallholder peasant households contribute over 95%, with a mere 5% coming from large-scale state and private commercial farms. Cereals and pulses are the most important staple foods in the country. Any change in the performance of these crops could therefore have a great impact on the total food supply in the country, as well as on agricultural GDP. Figure 3.3 portrays the annual changes and long-run trends in major food grain production per capita in Ethiopia. As is clearly depicted, the long-run curve (LR) of per capita food production is almost constant over the whole period of observation, implying stagnant food supply per person. Comparing trends between the pre- and post-reform period, the trends in the per capita food production index for the two periods are essentially the mirror image of each other.

Figure 3.3: Trends in Major Food Crop Production per capita (kg per person per year)

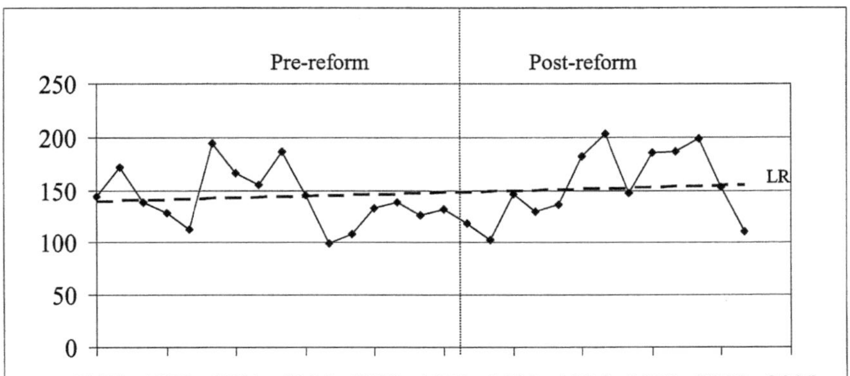

Source: Own computation.

Until the late 1950s, the country had been self-sufficient in staple food and was a net exporter of food grains. Afterwards, however, domestic production has at no time been able to meet even the basic minimum national food requirement (MEDaC, 1999). In the socialist period, between 1975 and 1991, food crop production was growing at a mere 1.2% while the population was growing at 2.64%, resulting in a downward trend in per capita production of food grain at a rate of -1.44%. With the exception of the early 1980s, the food production index declined continuously throughout the pre-reform period until the end of the regime. According to the survey report by MEDaC (1999), from 1980 to 1990, the area cultivated for cereals declined by 6.6% and that of pulses by 9.3%. As a result, cereal production grew by a mere 0.9%, whereas pulses fell by 1.7%. The problem was largely attributed to the misguided command policy of the socialist system, supplemented by recurrent drought and the civil war between the socialist government and the EPRDF (the then rebels). Consequently, total food imports, in the form of aid and commercial purchases, rose at a rate of about 20% per annum during that period.

After the policy reform, the overall performance of food crop production was appreciably positive between 1992 and 2001. The area of production increased by 63% and food production doubled. In this period, the trend was a positive upward one, and for the first nine years after the new government policy began to take effect, food production grew by about 34% (i.e., from an average of about 146 kg in 1993 to about 196 kg in 2001). In that period, the compound growth rate in food production was impressive, reaching 4.52% per annum (significant at 5%). As explained in the forthcoming chapters of this thesis, the main source of the improvement in food production in the period under observation was the improvement in cereals, mainly due to technological progress and increasing use of modern inputs (in fact only for some food crops), coupled with relatively good weather. Yet, crop yields in Ethiopia are about 2 to 3 less than in Asia developing and 5 to 6 times than in Germany and United States of America (see Table A3.2 in the Annex). However, progress suddenly collapsed in 2002/03 because of the customary incidence of drought and the failure of government policy to bring about sustainable production growth (explained in detail in Chapter 6).

As mentioned above, food production in Ethiopia is highly sensitive to climatic changes, as the production system is traditional and purely nature-dependent. As is clearly shown in the Figure 3.3, food production in Ethiopia is characterized by severe fluctuations that usually follow the fluctuations in the weather conditions. Particularly the years 1984/85, 1991/92 (due to civil war) and 2002/2003 were those in which the country experienced the lowest per capita domestic food production, and hence the worst hunger. In these particular years, 15% to 22% of the total population fell victim to the food crisis. The consequence has been hunger affecting millions and the death of thousands. In the last two years of observation, the food shortage was the most devastating the country has ever experienced, and the number of people stricken with hunger was unimaginably high, with as many 15 million affected, in other words, about a quarter of the total population. Overall, a minimum of 5% of the Ethiopian population needs food assistance every year. Assuming the minimum calorie requirement, which is equivalent to 225kg of food grain per person per year, the country has been able to satisfy on average merely 66% of the total food required in the post-reform period. The remaining 34% therefore represented a food deficit that had to be imported by purchase or in the form of aid. In general, food production per capita remained stagnant, and was characterized by severe fluctuations during the whole observation period, both before and after the reform. Consequently, an increasing share of the population, particularly in marginal and drier areas, was tormented by hunger every year (for the number of people affected by disaster and the annual volume of food aid, see Table A3.3 in the Annex).

Because production was inadequate to satisfy domestic demand, the nation was obliged to allocate, on average, close to 10% of the value of imports (which is about 25% of the value of exports) to importing food items in the post-reform period. The most terrible problem is that because millions of poor people cannot afford to purchase food from the market, resorting to food aid is the only option available to fill the food gap. Especially people in marginal and drought-prone areas, notably in wide areas of the eastern and northern part of the country, will continue consistently to seek food aid unless a special development strategy is designed for them. Thus, for decades in Ethiopia the gap created by domestic production shortfalls has been largely met through bilateral and multilateral food aid. The annual volume of food aid ranges from 2.3% to 26% of total domestic grain production over the period 1983-2003.

In conclusion, then, the economic reforms were unable to help the country escape from the problem of scant domestic food supply in relation to rising demand. While the supply of agriculture has increased considerably, the crucial problem of food insecurity in the country remains a challenge for policy. This greatly affects the potential for growth and stability of the economy for the coming decades.

3.3.3 Agriculture's Contribution to Exports and Changes in the Performance of the External Sector after the Economic Reforms

Major expected outcomes of the trade policy reform include increasing and diversifying exports, rendering the domestic economy more efficient and improving its competitiveness in the world market. The reform in Ethiopia was also expected to result in a significant increase in the volume and diversity of exports and increased export earnings, which would ultimately bring about an improved trade and current account balance. Obviously, almost all export earnings from goods come from agricultural products, which account for over 85% of the total.[7] The relative importance of the five major agricultural products, coffee, hides and skins (including leather & leather products), oil seeds, pulses, and fruits and vegetables remained almost the same. Altogether, they accounted

7 Detailed statistics for exports and imports for several years are provided in the Annual Reports and Quarterly Bulletins of the National Bank of Ethiopia. Since changes in the service exports were steady throughout the period under observation, much of the change in export earnings was due to changes in merchandise exports.

on average for 75-80% of total export earnings. The share of non-agricultural goods in export earnings was negligible and was persistently below 15% in all periods. New export items in terms of agricultural and/or industrial products were virtually absent. In both pre-and post-reform periods, coffee stands out among all the sources of export earnings, contributing about 60%, except from 2000 onwards, when its share shrank to about 40% owing to fall in the price of coffee on the world market. On the other hand, the percentage contributions of "chat" and oil crops started to rise consistently, starting in 1996, and an impressive resurgence was recorded in their export volume. So, although no significant change was noticeable in the structure of export goods, the importance of the latter products was mounting. Nevertheless, this suggests that the reform of foreign trade had little impact on the structure of export goods. In other words, the export items of the country in general could not be plausibly differentiated throughout the observation period (Table 3.2).

Table 3.2: Export Structure and Performance: Percentage Contribution of Agricultural Products to the Total Value of Export Goods

Period	Coffee	Hides & Skins	Oil seeds	Pulses	Chat	Other Agri.	All Agri.	All Non-Agri.	Average Total value (birr mns)
1987-1990	60.57	15.33	1.58	2.39	2.37	6.41	88.65	11.35	818.30
1991-1995	54.08	15.24	1.15	1.78	5.16	6.39	83.80	16.20	1218.70
1996-1999	63.32	9.13	4.62	2.56	7.78	1.87	89.28	10.72	3569.40
2000-2003	40.88	11.68	7.55	3.83	12.92	5.29	84.15	15.85	3957.80

Source: Computed from National Bank Quarterly Bulletin (2002/2003).

Indeed, the volume of exports of goods rose sharply, as expected, by an average of 34% between 1992/93 and 2002/03 after the reform. The total value of exports earnings increased more than threefold in the post-reform period compared to the pre-reform period. Total export earnings increased from 10% of GDP in the pre-reform period to more than 15% after the reform. Compared to the years before the reform, the annual average nominal value of foreign earnings from the export of goods consistently increased, and in the decade from 2000 it was more than tripled. Besides the volume of exports, export earnings by and large depend on the behavior of export volumes and the unit values of the traded item. In this regard, it is worth noting the conspicuous instability in the export volumes because of the inherent feature of agricultural products, i.e. the fact that they are susceptible to weather changes and price fluctuations in the world and domestic markets. As coffee constitutes the bulk of Ethiopia's export earnings, any shock in this product is automatically transmitted to the total export earnings. Unfortunately, coffee suffers from frequent shocks in world prices, which has a profound effect on domestic coffee farmers and on the export earnings of the country. Consequently, the total value of earnings from coffee was found to be unstable due to high variability in its unit price in the world market, and this in turn increases the variability in its volume of production and supply.

Similar to exports, the structure of imports of good and services was more or less the same throughout the observation period, and no dramatic change was noticed after the reform. On average, imports of consumption goods accounted for about 30%, capital goods 35% and raw and semi-finished materials (mainly fuel and fertilizer) 18% of the total value of imports. Imports of the latter represent a very small proportion of imports. However, the total value of imports increased radically (fivefold) in the post-reform period, over and above the increase in exports. This may have been due to the removal of quotas and cuts in tariff rates, which in turn encouraged an increase in the consumption of imported goods and imports of raw materials and capital goods for investment.

Table 3.3: External Balance and Debt Status of the Economy

Year	Trade Balance (birr mns)	CAB (birr mns)	External Debt to GDP Ratio (%)	Debt Service Ratio
1988	-1486.6	-1163.8	41.3	43.5
1989	-1192.2	-706.6	41.2	39.2
1990	-1067.3	-623.5	43.1	58.9
1991	-1558.2	-1206.3	39.1	69.9
1992	-1491.7	-810.6	31.5	82.5
1993	-2668.8	-1646.9	70.4	53.0
1994	-3321.1	-1713.6	90.8	56.9
1995	-3711.2	-1433.3	81.8	36.7
1996	-6255.5	-3860.5	71.4	34.5
1997	-4613.7	-2282.6	63.9	39.8
1998	-5196.4	-2360.1	62.0	19.2
1999	-8064.7	-5270.4	65.2	22.9
2000	-9158.0	-5007.8	86.1	27.8
2001	-9287.9	-5483.3	89.0	21.1
2002	-11020.0	na	103.4	14.7

Source: Compiled from Annual Reports of National Bank of Ethiopia.

With relatively smaller, though expanding, exports and a sizeable increase in imports, the trade balance was consistently and increasingly negative in the observed period after the reform, owing to the disproportionately larger imports compared to the marginally increasing exports. The trade deficit widened from an initial level of 1067 million birr in 1990 to about 3036 million birr in 2003, an increase of nearly three hundred percent despite the favorable increase in the volume of exports (Table 3.3). Consequently, despite trade liberalization, the current account balance (CAB) deteriorated further in this period. The chronic problem of external deficit remained a critical problem of the external sector. Thus, the Ethiopian experience did not bear out initial optimism regarding the positive impact liberalization would have on the trade and current account balances. In general, growth in export earnings was sluggish and fluctuating, whereas imports grew consistently fast, which ultimately resulted in an increasing deficit in the external balances. In addition, as Table 3.3 indicates, the ratio of external debt to GDP was continuously increasing, and in 2002 it exceeded 100%, i.e., the debt became greater than GDP in absolute terms. On the other hand, the debt service ratio was declining severely. This implies that the country's ability to pay its debt from its export earnings was increasingly weakening.

The fact that Ethiopia continues to rely on exports of the same goods (primary) and imports of the same goods and services as it did more than 50 years ago indicates the failure of the policies of all government regimes to bring about diversified economic growth; it also indicates that the country's economy is static. Moreover, it implies that the country produces and consumes the same goods, using more or less the same level of technology. The backwardness of the country's agriculture testifies to the fact that little innovation has been introduced into to this critical economic sector. Because the prices and the value-added of primary goods are far lower than that of processed goods, and decline over time, the country should diversify its export sector by all possible means. The challenge to policy-makers and planners is, therefore, how to initiate and instill dynamic comparative advantage into production processes.

3.3.4 Contribution of Agriculture as a Supplier of Surpluses to Other Sectors of the Economy

The role of agriculture in supplying surpluses and its consequent impact on the growth of other sectors, particularly industry, is best explained by exploring growth linkages using economic models such as *Leontief Input-Output* model, or more advanced models like *Social Accounting Matrices* or *Computable General Equilibrium* modeling, or econometrics models (*simultaneous functions*). However, this section provides only a brief description of changes in share of the labor force, and import intensity of raw materials, representing the supply by agriculture of labor and raw materials respectively.

3.3.4.1 Labor Supply and Challenges Relating to Changes in Labor Force Structure

Transformation of the structure of the labor force is a prerequisite in the development process for industrialization to occur. Given that agriculture is a labor reservoir sector, in which more than 85% of the Ethiopian population is engaged, it may be expected to release some labor to the non-agricultural sector (particularly to industry), provided that the latter grows faster as a result of policy reform and attracts agricultural labor by paying a wage rate that is at least higher than the average farm income. In this review, data on labor movements from one sector to another was not available. However, a careful assessment of growth trends in the labor force in the agricultural and non-agricultural sectors can highlight whether or not labor transfer may have been occurring between these broadly classified sectors.

In the post-reform period (1993-2000), the medium and large-scale industries grew at an incredible rate (15.4%). However, the change in their share of the labor force was insignificant, and the agricultural sector still embraced the vast majority of the labor force. Table 3.4 summarizes trends in labor force changes in the agricultural and non-agricultural sectors in the period in question. In twenty-five years (1975-2000), the share of the non-agricultural labor force only grew from 10% to 15%, leaving the bulk of the labor force to the agricultural sector. The role of the medium and large manufacturing sector in employment was very limited owing to its relatively high capital intensity (MEDaC, 1999). In the period from 1975 to 2000, this sector employed an average of only about 0.35% of the total labor force. The growth rate of employment in this sector was dismal and not promising. During the socialist period, it grew to about 1.95%, but worsened after the reform, falling to a rate of about 1.74% despite the increasing number of newly-established private manufacturing industries.

Table 3.4: Structure of Labor Force in the Agricultural and Non-agricultural Sectors (in millions)

Year	Labor in Agriculture (a)	Labor in Non-Agriculture (b)*	Labor in LM@ Industries (c)	Total Labor Force (a + b)	Total Population
1975	13.7 (90%)	1.47 (10%)	0.059 (0.39%)	15.159	34.40
1991	19.8 (86%)	3.18(14%)	0.076 (0.33%)	23.031	52.26
2000	22.7(85%)	4.10 (15%)	0.950 (0.36%)	26.723	63.50
Growth (Rate %):					
1975-1991	2.52	5.34	1.95	2.83	2.90
1993-2000	1.58	3.30	1.74	1.84	2.57
1975-2000	2.26	4.83	1.34	2.57	2.64

Source: computed from the CSA Annual data.

Note:@ indicates large and medium industries; * Includes labor force in all categories of industrial and service sectors; in the year 2000, total labor force is less than 50% of the total population.

In relative terms, the highest concentration of employment was observed in the food and beverage and textile industries, which consistently accounted for around 60% of the total employment in medium and large industries. The dominance of these two industries is a pattern that is familiar in many developing countries. More importantly, the slight positive increase in the non-agricultural labor force was in fact attributed to the contribution of the small-scale and cottage industries, and the service sector. According to a 1995/96 survey (MEDaC, 1999), over two million people were engaged in these industries, 70% of whom were unpaid family labor. Production of goods and services depended entirely on domestic raw materials, with negligible imports, and food items were the dominant product produced, accounting for over 60% of total gross values.

Growth of the non-agricultural labor force was rather slower (3.3%) after the reform compared to the pre-reform period (5.34%), although it grew relatively faster than the agricultural labor force (1.6%) and the total labor force (1.84%). The slower growth in non-agriculture labor in the post-reform period could be ascribed to the decline in government spending on industry (Figure 3.1), and the increased capital intensification of private industries despite the increase in the number of new establishments (Figure 3.4). In this regard, the rate of labor force growth in the non-agricultural sector by comparison to that of the agricultural sector was somewhat better in the pre-reform period, at 5.34% and 2.52% respectively.

In confirmation of the general growth trend of the development process, in the whole period from 1975 to 2000, the non-agricultural labor force was growing more rapidly (4.83%) than that of agriculture (2.26%) or the total labor force (2.57%). In the same period, the population growth rate was 2.64%, which was also much lower than labor force growth in the non-agricultural sector. Generally, the non-agricultural labor force grew faster than the agricultural labor force or the total labor force. This provides an indication of the way ahead to structural transformation, even though a considerable number of years will be required to reach a turning point (where the rate of agricultural labor growth becomes negative). This is because the higher agriculture's initial share in the total labor force and the faster population growth rate, the longer it takes to reach the turning point (Eleni et al., 1999).

3.3.4.2 Use of Raw Materials

Based on the low level of labor employment, one can anticipate that the medium and large industries will be capital-intensive. Import intensity (computed as the ratio of costs of imported raw materials to total cost of raw materials consumed) highlights the relative importance of imported raw materials used in the manufacturing sectors. Accordingly, the modern industries were found to be heavily import-dependent with an aggregate import intensity of about 0.47. In other words, for more than the last 20 years, raw materials equivalent to about 50% of the total inputs consumed in the process of production were imported. In Figure 3.4, the curve shows a slight decline in the import intensity of raw materials in the LM industries in the pre-reform period, perhaps due to the import substitution policy then in place. The curve reached its minimum level in early 1990, the year in which the number of establishments was also at its lowest. In the post-reform period, import intensity started to rise with the increasing number of establishments (with small fluctuations), averaging about 0.48. The largest import-dependent industry was the metal industry, with an import intensity of about 0.95, followed by the chemicals and plastics industry, with an import intensity of over 0.80 in 2000. The least import-dependent industry was food and beverages (the largest industry in the economy), followed by the leather and textile industries, with average import intensities of 0.23, 0.24 and 0.32 respectively in the years 1996 to 2000.

Figure 3. 4: Number of Establishments and Raw Material Import Intensity of Medium and Large Industries

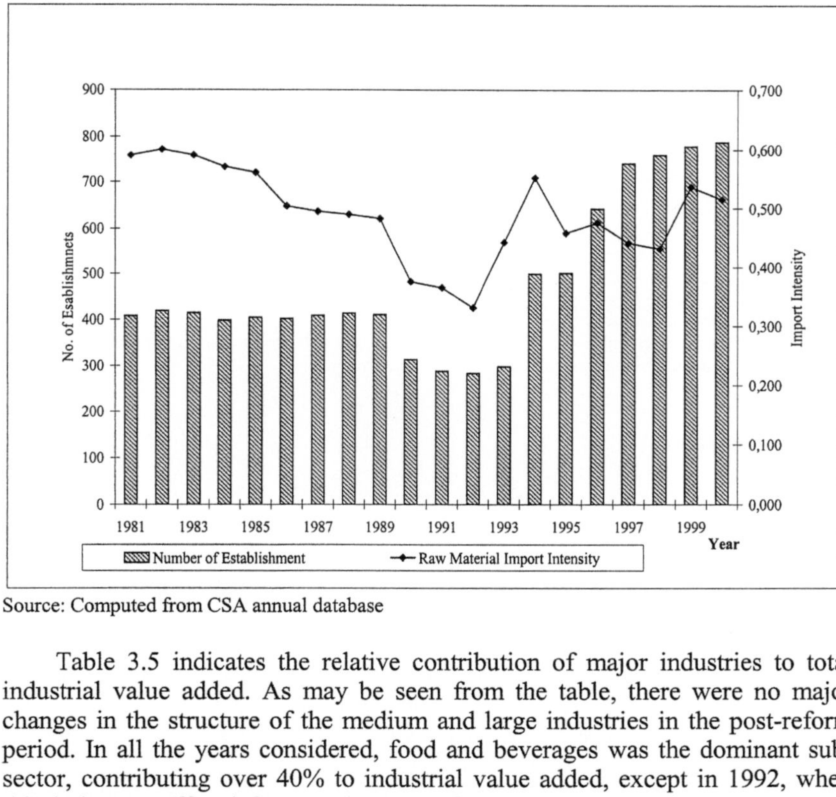

Source: Computed from CSA annual database

Table 3.5 indicates the relative contribution of major industries to total industrial value added. As may be seen from the table, there were no major changes in the structure of the medium and large industries in the post-reform period. In all the years considered, food and beverages was the dominant sub-sector, contributing over 40% to industrial value added, except in 1992, when this industry suffered from a shortage of domestic raw materials because of political disruption. Until 1996, a textile were the second most important industry, contributing over 20% to total industrial value added. Since 1996, metal has taken over in second place, increasing its contribution from its lowest level of 3% in 1982 to about 12% in 2000. The trend of substitution of the textile industry by the metal industry worsens the situation as regards self-reliance in domestic resource use and employment. This is because the textile factories use local resources (mainly agricultural products) more intensively and are relatively more labor-intensive, whereas the metal industry is more import-dependent and capital-intensive. If such a paradigm continues to increase, therefore, it will directly and profoundly depress the process of structural transformation of the economy by weakening the forward linkages between the agriculture and the industrial sector. In other words, demand for domestic agricultural products as raw materials by domestic industries will fall, and domestic industries will benefit less and less from growth in the agricultural sector.

Table 3.5: Distribution of Value Added by Major Medium and Large Industrial Groups (in percent)

Major Group of Industries (LM)	Year 1982	1992	1996	2000
Food & Beverages	53.00	30.70	41.16	44.10
Textiles and Apparel	21.82	24.25	10.78	6.40
Leather and Shoes	2.80	5.51	9.15	4.40
Metal	3.68	2.50	7.87	12.30
Other*	18.70	37.04	31.04	35.80
IC/VA	0.09	0.21	0.14	0.19

Source: computed from series of CSA annual reports
Note: Other* includes chemicals, rubber, plastics, tobacco, wood products, and non-metals; IC/VA is incremental capital to average value added at factor cost ratio.

3.4 Conclusion

Three distinctly different political regimes, each with their own singular development policies and strategies, have been experienced in Ethiopia. In the pre-reform period, the feudal-bourgeois regime practiced both outward and inward-oriented development policies, while the socialist regime was strictly inward-oriented. In both of these earlier regimes, the main focus was on development of the industrial sector, while agriculture received less attention. Coupled with periodic drought and civil war, the misguided economic policies of the feudal-bourgeois regime and the socialist dictatorship severely harmed all sectors and more seriously damaged the dominant sector of the economy, agriculture. Agricultural production stagnated and food production per capita fell drastically, resulting in massive hunger, leading the government to seek food aid and increasingly allocate foreign exchange to pay for food imports. The ultimate result was a gradual reduction in the living standard of Ethiopian society as a whole.

Since policy reforms were launched by the current government in the early 1990s, development policies and strategies have been designed in such away that agriculture is given priority and plays the leading role in overall economic development. While slight positive trends may be seen in the growth of some sectors, no appreciable difference was achieved in terms of the performance of the per capita indicators for the economy as a whole. Growth in GDP and food production was too sluggish to cope with the growing population, and thus per capita GDP and per capita food production remained stagnant. In addition, no significant difference was observed in the structure of the economy; there is still a long way to go to achieve this. The development impact of policy reform in Ethiopia at the macro level is therefore unclear as yet.

Since the launch of the policy reform, the performance of the agricultural sector at the macro level has been not satisfactory, and poorer than expected given the great emphasis place on agriculture in the government's development strategy. Because sustainable use of modern inputs (such as improved seeds and fertilizers) has been lacking in agriculture, the sector's performance has generally been poorer than it was projected by the government. Food production particularly has been unreliable and is still far below the food requirement of the nation, and the contribution of the sector to the economy as a whole has been inconsequential. The principal problems emanate from the prevailing problem of unreliable climatic conditions and unstable prices for farm products. This clearly indicates a need for government policy to promote irrigation technology. Furthermore, although the free market is presumed to guide and stabilize the economy, instability in output prices has made small farmers increasingly vulnerable to incalculable risks. This critical problem could ultimately create a disincentive to the use of improved technologies and hence lead to unsustainable agricultural development.

Apart from an increase in the volume of agricultural exports (but without diversification), it has not been possible to improve the external balances. Rather, the trade and current account balances have tended to deteriorate since the economic reform. The main reason for this may be that the country was increasingly importing much larger amount of goods (particularly consumption goods) than it was exporting, and partly because of the declining trend in world prices for primary goods (particularly for coffee, the countries main export). The other disappointing feature of the economy is that both agriculture and domestic manufacturing industries were increasingly depending on imported raw materials, while the linkages between the two were tending to loosen. From a development theory perspective, these conditions are likely to prove unsustainable. In conclusion, both the Ethiopian economy in general and the agricultural sector in particular face a number of multifaceted problems, and thus there is a need to review current development policies and strategies.

4 Analysis of Dynamics in Sources of Agricultural Output Growth at Farm Household Level

Because of the depressive effects of economic central planning under the Socialist regime and the absence of improved technologies, farm productivity in Ethiopia was stagnant and growth in agricultural production was sluggish in this period (see Chapter 3 for detailed discussion). Consequently, there was little market surplus and linkages between the agricultural and the industrial sectors. In Chapter 2, we examined in detail how structural and institutional reforms influence resource use and growth in output. The broader theoretical concepts of different development approaches aimed at accelerating agricultural growth, supported by empirical evidence, were also highlighted.

It is clear, therefore, that transforming economic policy and institutions from a centrally planned to a market system creates a new environment that may provide the basis for optimal production and consumption decisions and new economic structures in Ethiopia. Reforms such as market liberalization, introduction of improved technologies, and better access to information are expected to encourage farmers to enhance productivity and growth through better use of resources and increased production potentials.

Methods of measuring performance can be applied at both micro and macro level. For instance, changes in technical efficiency (TE), technological progress and productivity growth are best measured at micro level, whereas price stability, demand for modern inputs and supply responses can best be measured at macro level. The latter are presented in *Chapter 6*. Assuming that there are no major changes in natural factors during the study period, a change in growth trends of farm output and agricultural productivity of Ethiopian smallholders could be linked to the impact of the reforms and institutional changes undertaken by the government. Producers may respond to policy reform incentives by increasing their use of resources, adopting improved technologies or improving resource use efficiency. Before embarking on analysis of the impact of the reforms on growth in productivity and output, it is useful to outline in brief the classical sources of output growth.

4.1 Theory Relating to Sources of Output Growth and the Role of Institutions

In this section, the theory relating to major sources of output growth is discussed in brief to establish a link between the theoretical and empirical analytical approaches. Generally speaking, there are three conventional sources of output growth: expansion/intensification of resource use, improving utilization of the available resources (efficiency improvement) and changes relating to technological progress. The latter two result in growth in total factor productivity. These sources of growth are further explained using a single input-output relationship as portrayed in Figure 4.1. In this Figure, F_1 represents a production frontier in traditional production and F_2 represents a production frontier using improved technologies, and L_1 and L_2 are the iso-profit line on the respective production frontiers.

Figure 4.1: Sources of Output Growth

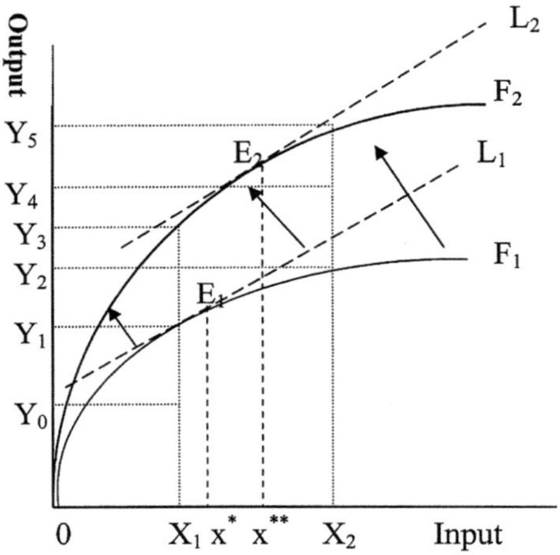

Source: Own demonstration

4.1.1 Expansion or Intensification of Resource Use

Production can be increased from increased use of inputs. This approach is also called horizontal expansion, which can be measured by the movement from a given point. As indicated in Figure 4.1, using traditional practices, by increasing input use from X_1 to X_2, output can be increased from Y_1 to Y_2 on the same production function F_1 under technically perfectly efficient conditions. The principal reason for producers to use more inputs (assuming idle resources are available) is an increase in output prices or a decrease in input prices, or both. For small farm households, an increase in demand for food may have a greater influence than price changes. This source of economic growth, however, has little applicability at global or national level due to current socio-economic and environmental restrictions on resources. In Ethiopia too, this option is hardly feasible, as farmers do not have the capital to buy inputs such as machinery, expansion of land use, on the other hand, may be possible in areas where land is an idle resource given the existing technology.

4.1.2 Improving Efficiency

As resources are generally scarce, the concept of efficiency has become the core of economics. Improving the efficiency of a firm is bound up with possibilities for minimizing costs by combining inputs depending on relative prices (allocative efficiency), or maximizing the level of production from the given set of inputs (technical efficiency). The compound function of these two efficiencies gives us economic efficiency. The other method of improving efficiency is by identifying the most efficient scale of production (scale efficiency). In the following section we will discuss allocative and technical efficiency in more detail; scale efficiency is explained in Section 4.5.3 using the empirical results of the study.

4.1.2.1 Allocative Efficiency

Allocative efficiency is also referred to as price efficiency. It shows one's ability to maximize profit or minimize cost. This is achieved at a production level where the marginal revenue product of the factors of production is equal to their marginal cost; or simply, when the unit price of inputs is equal to their respective marginal revenues. A deviation from equality implies that the resources are inefficiently utilized, or simply shows the firm's inability to use inputs in optimum proportion given their respective prices (Coelli, Rao and Battese, 1998). This means that allocative inefficiency raises the average cost of production by utilizing combinations other than the cost-minimizing combination of inputs, implying a need to reallocate the inputs to increase profit. The inefficiencies are regarded as consequences of distortions in prices and quantities of inputs used.

The underlying assumptions for allocative efficiency to prevail include the existence of a perfectly competitive market (in other words, where prices are a true measure of resource scarcity), the availability of full information to all economic actors, and the assumption that a producer's goal is profit maximization. Based on Adam Smith's *invisible hand* concept, perfect market prices lead to optimal allocation of resources. In Figure 4.1, the tangent points (E_1 and E_2) of the iso-profit lines and the production frontiers indicate the level of production using x^* and x^{**} levels of inputs respectively.

Although neo-classical microeconomic theory focuses on the relevance of allocative efficiency for growth, the underlying assumptions are said to be unrealistic and often remain controversial in an environment of full of complexities and externalities. For instance, when prices are distorted by governments or monopolies, or when goods remain outside the market system because of poor market development, the role of prices in resource allocation is greatly impaired (Vogit and Uvarovsky, 2001). In this respect, many have tested and rejected the assumption that farmers maximize profit (Junanker, 1989; Dean and Perlman, 1998). They abandon the profit-maximization assumption altogether and assume that neither firms nor individuals necessarily maximize profit. Because of the unrealistic assumptions, Farrell (1957) also criticizes allocative efficiency measures as both unstable and dubious. Furthermore, Leibenstein (1966) goes as far as to argue that the impairment to economic growth due to allocative inefficiency is negligible, and emphasizes instead the importance of improving technical efficiency. In general, the arguments in favor of technical efficiency are based on the fact that it is relatively less heavily dependent upon the aforementioned assumptions and that it contributes more to economic growth than allocative efficiency does (Schultz, 1978). For this reason, we have not considered allocative efficiency in greater detail in our analyses.

4.1.2.2 Technical Efficiency

Technical efficiency (hereafter TE), also referred to as managerial or institutional efficiency, can be defined in many different ways.[8] For the purposes of this study, the definition of TE presented by Battese and Coelli (1995) is used. In their seminal works on efficiency, TE is defined in terms of the production function that relates various levels of inputs and outputs: it is the ratio of the actual production of a firm (farmer) to the potential production given by the hypothetical production function. This implies that as the ratio approaches unity, the firm is approaching its maximum TE. A technically perfectly efficient unit operates on its frontier production function to obtain the maximum possible output that is feasible given its current technology. Leibenstein (1966) described technical inefficiency as:

> *"Firms and economies do not operate on the outer-bound production possibility surface consistent with their resources, and any combination of input and output is possible under the curve. But, they actually operate on a production surface that is well within that outer bound. This means that for a variety of reasons people and organizations normally work neither as hard nor as effectively as they could."*

Technical inefficiency effects define the degree to which firms fail to reach the frontier (potential) production (Lundvall and Battese, 1998). Thus, if farmers operate below the frontier (i.e., if they are inefficient), there could be scope for increasing output by improving utilization of the same resources and without incurring significant investment costs. On the production curve (Figure 4.1), this can be indicated as a movement of the output level from Y_0 to Y_1 on frontier F_1

---

[8] For details of the various definitions see, e.g. Färe, Grosskopf and Lovell, 1985.

or from Y_4 to Y_5 on F_2. Moving the output level upwards means increasing productivity by improving TE. The approach of increasing productivity by improving efficiency of resource use requires the removal or mitigation of socio-economic and institutional constraints.

To a large extent, TE avoids many of the criticisms leveled at more general efficiency concepts. In particular, measures of TE rely less heavily on the assumption of a perfectly competitive market, perfect knowledge and profit maximization objectives. Colman and Young (1990) also underlines that TE refers only to the physical characteristics of the production process. It can therefore be taken to be a universal goal in that it is applicable in any economic system.

4.1.3 Production Technology

Technological progress, also called the transformation approach, is the other most important source of growth in a traditional economy like Ethiopia. According to Colman and Young (1990), technology is defined as a stock of available techniques or a state of knowledge concerning the relationship between inputs and given physical outputs. They further describe technological change as an improvement in state of knowledge that manifests itself in the enhancement of production possibilities. That is to say, technological change results in a shift of the production function over some range in such a way that more output can be produced with the same quantity of inputs or the same output can be produced with a smaller quantity of inputs. The impact of technological change on production growth can be illustrated in different ways: an outward shift in total production or production possibility frontier, or a downward shift in the isoquant curve.[9] Using the input-output relationship in Figure 4.1, the shift from F_1 to F_2 is the result of technological progress. Consequently, using a level of inputs of, say, X_1, output is increased from Y_1 to Y_3; or, possibly, using X_2 level of inputs, output can be raised from Y_2 to Y_5 under efficient conditions.

Assuming constant factor prices, technological change is categorized as *neutral* or *non-neutral (biased)* depending on induced changes in optimal factor proportions (e.g., capital-labor ratio, K/L). A technological change is said to be neutral if it induces no change in factor proportions, i.e., if it remains constant. As explained by Thirtle and Ruttan (1987), assuming that for an individual firm prices of inputs are given and factor endowments are fixed in the short term,

---

[9] Detailed explanations and graphic presentations of the impact of technological changes are provided in Colman and Young (1990), pp. 54-56.

neutral technological change can be more appropriately defined, with the *given factor proportions*, as raising the marginal product of inputs by the same proportions (i.e., the marginal product of labor increases by the same proportion as the marginal product of capital or land). On the other hand, technological change is biased in the sense that, at constant factor prices, it induces a change in optimal factor proportions. In such a case, the same optimum level of output can be produced using less of one and more of another input. For instance, labor-saving technological change makes labor more plentiful relative to the other inputs and results in a fall in the marginal product of labor relative to other inputs; land-augmenting technologies increase the marginal product of land relative to other inputs (Thirtle and Ruttan, 1987).

Technological change is of paramount importance particularly where there is limited potential for horizontal expansion and improvements in technical efficiency to boost growth. In farm production, use of improved seeds, modern machineries, improved application of chemical fertilizer and better agronomic practices are among the major technical changes that result in technological progress. The possibility of implementing such innovation, however, is crucially dependent on well-functioning credit and market institutions, and efficient policy instruments in general.

4.2 Measuring Changes in Productivity and the Relative Contribution of Sources of Output Growth

Measuring agricultural performance and its efficacy is one of the most important economic research objectives of this study. An important economic presumption throughout this study is that institutional failures and incompleteness are important sources of real divergences from neoclassical production as described by production function; and this includes any type of inefficiency. Measuring productivity growth in particular has attracted many economists and policy makers because a change in agricultural productivity determines the potential for output growth and the competitiveness of the product in the local and world market. Productivity index provides a number or score for output in relation to inputs used. This implies that, in essence, measuring productivity growth is about measuring how the ratio of output to input changes over time, which essentially tells us the level of input utilization over time. There are many approaches to measure changes in productivity. Here, they are broadly classified as conventional and decomposition approaches.

4.2.1 Conventional Approaches and their Shortcomings

The most common and easiest way is to compute partial productivity, whereby total output is divided by input, say, labor or land. The coefficient of such a computation is referred to as the average productivity of the input under consideration. Generally, simple indicators of productivity measurement are labor or land or capital productivity. Due to its limitation to only one input, it appears to neglect the effects of other factors, which is acceptable only under a *ceteris paribus* clause, or in instances where individual factors are insignificant in the production process (Dyckhoff and Allen, 1997). Partial productivity also ignores input substitution. For instance, an increase in the use of labor and capital can increase land productivity, which in turn results in changes in the land-labor and land-capital ratio. Hence, partial productivity does not provide appropriate information about actual changes in the output-input relationship.

The best and most robust measure of gains in productivity is "total factor productivity", or TFP, defined as total output produced per total input used. It can be also technically explained as the combined effect of changes in technological progress and technical efficiency. Traditionally, there are different methods of computing TFP. The most commonly used are the index-number approach and measuring a representative function of technology (e.g., a distance function, cost or revenue function and profit function). The index-number approach in particular is most frequently used (Diewert, 1976) if detailed data on each component of input and output are available. Obtaining detailed data on smallholders, however, is difficult. Computation is also tedious and time-consuming. Moreover, these traditional measures of productivity do not distinguish between the components of the TFP. They assume technological change as the unique source of growth in productivity, assuming all producers are perfectly efficient; hence, TFP growth is often used synonymously with technological progress. On the other hand, the effect of TE improvements on productivity growth is often overlooked. Meanwhile, failure to take account of changes in TE when measuring TFP growth gives biased estimates of technological progress. For instance, a high rate of technological progress can coexist with deteriorating efficiency and vice versa.

For these reasons, a special model that can decompose productivity and output changes into their components is essential. Models such as programming, deterministic statistical frontier, parametric frontier, stochastic frontier production function approaches, etc. have all been suggested. But Pitt and Lee (1981) proved that the programming and the deterministic statistical approaches do not allow for random shocks in the production process, and thus a few extreme observations can exaggerate the maximum possible output. In addition, they do not account for statistical noise and assume that all deviations from the frontier are due to inefficiency. The stochastic frontier approach has been found to be superior and avoid the kinds of problem described above. Alternative approaches to study impacts of interventions are summarized in Box 4.1.

4.2.2 A Growth Decomposition Approach: Application of the Stochastic Frontier Production Function

The stochastic frontier production function approach was first developed by Farrell (1957), and has been expanded by various authors (Aigner, Lovell and Schmidt, 1977; Battese and Coelli, 1995) to estimate production function and efficiency. A frontier production function has increasingly been used as a methodological tool. The basic concept behind this model is that it is originated from "best practice techniques" and captures two types of error, the conventional random component (white noise), and errors due to the effects of inefficiency as the source of the deviation from the frontier (Battese, 1992). As it also allows observation to fall above the production function, the model is less sensitive to the influence of outliers on the technical inefficiency estimates. Therefore, it is more closely related to the theoretical definition of production function which relates to the maximum attainable output from a given sets of inputs.

The stochastic frontier approach has great advantages in modeling output growth in terms of evaluating the effects of policy changes: first, it allows the filtering out of the random effects typical for traditional agriculture, such as weather changes and measurement errors. Second, estimates of a firm's TE can be obtained by comparing observed and predicted output; and third, it allows a functional specification for production technology and a test of its statistical properties. The stochastic model considers the various techniques of application of inputs that are assumed to influence output. Output growth can therefore conveniently be decomposed into its components – TE, technological change and input effects – which are of course policy and institutional variables (Kalirajan, et al., 1996). The relative importance of these components can also be estimated. In this context, the stochastic frontier approach employed further benefited from the modeling framework discussed in Fan (1991), Salim and Kalirajan (1999) and Voigt and Uvarovsky (2001).

The methodology for estimating the impact of changes due to the economic and institutional reforms on productivity (technological and efficiency changes) and then on output growth is depicted in Figure 4.2. To make the concept easier to understand, a simple deterministic production function is used.

Assuming two periods, a producer faces production frontier F_1 and F_2 in period I and period II respectively. If production were perfectly efficient, outputs would be Y_1 in period I and $Y`_2$ in period II, using $X_1$ and $X_2$ levels of input respectively. If the producer continues to use the same technology throughout the periods, but increases use of inputs from $X_1$ to $X_2$, then the change in input use would contribute to output increase from $Y_1$ to $Y_1$` on F_1 (assuming perfectly efficient production). However, the producer actually realizes a lower level of output, say Y_0 in period I and Y_2 in period II, owing to production inefficiency.

Improvement in TE reflects an individual's ability to improve production with the given inputs and technology by adopting best practice (techniques). Output growth that occurs through improvement in TE originates from the movement of the producer's actual output towards the maximum possible output or frontier, given the same technology (from Y_0 to Y_1 on F_1, or fromY_2 to $Y`_2$ on $F_2$). TE improvement, therefore, is the decrease in distance between the producers' realized output and the potential output over time. If, in the figure, we let $Y_1$-$Y_0$ (= $E_1$) be technical inefficiency in period I, and $Y_2$`-$Y_2$ (= E_2) in period II, the difference (E_2-E_1 = ΔE) shows the change in efficiency in the two periods. It shows a deterioration in TE if $\Delta E > 0$ or an improvement in TE if $\Delta E < 0$, and otherwise constant TE if $\Delta E = 0$.

On the other hand, technological change reflects the impact of using new/improved technologies on output growth measured by its impact on the distance between frontiers F_1 and F_2 using either X_1 or X_2 level of input. Under normal economic conditions, the index of technological change is positive. Accordingly, output from a given level of input increases by the degree of shift in the production frontier, for example, from $Y`_1$ to $Y`_2$ using the X_2 input set. To make it clearer, sources of output growth can be disaggregated from the Figure. Note that Y_o and Y_2 are the actual output in periods I and II respectively and the difference between them can be considered as actual output growth. The partial contribution of changes in TE, technological change and TFP can be algebraically derived as follows:

$$
\begin{aligned}
\textit{Output growth} &= Y_2 - Y_0 = A+B+C \\
&= (Y_1 - Y_0) + (Y_1` - Y_1) + (Y_2 - Y_1`) \\
&= (Y_1` - Y_1) + (Y_2` - Y_1`) + (Y_1 - Y_0) - (Y_2` - Y_2) \\
&= (Y_1` - Y_1) + (Y_2` - Y_1`) + E_1 - E_2 \\
&= (Y_1` - Y_1) + (Y_2` - Y_1`) + \Delta E \\
&= \textit{input growth} + \textit{technological change} + \textit{change in TE} \\
&= \textit{input growth} + \textit{TFP growth}
\end{aligned}
$$

Figure 4.2: Decomposing Effects of Input, Technological Progress and Improvement in Technical Efficiency on Output Growth

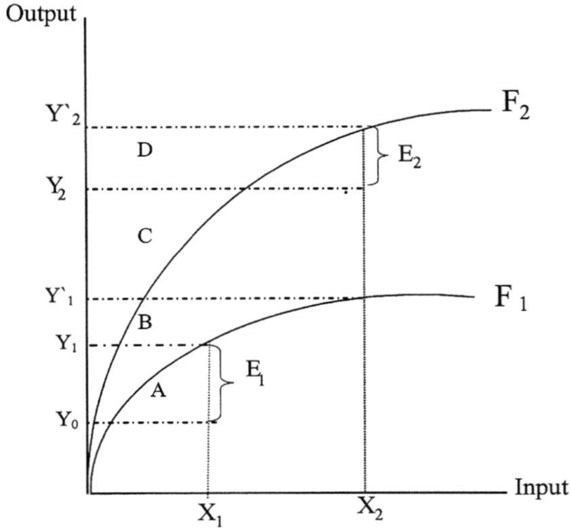

Sources: Modified from Fan (1991) and Salim and Kalirajan (1999)

4.2.2.1 Time-Varying Inefficiency Model

Figure 4.3 shows the principle of the stochastic frontier model with two distinct production functions for different technological transformations. Production function F_1 represents the conditional mean output level from the set of inputs used given initial technological conditions, and F_2 represents the conditional output from the set of inputs in the second period assuming technological change over the given period of time. Such technological change shifts the production function as depicted, which results in a new intercept and possibly new slopes as well. The observed production differs from the deterministic frontier production, $y = (X_i, \beta)$, due to technical inefficiency u_i and the random effects of v_i. Because the random error v_i can be positive (i.e. lies above the deterministic frontier) or negative (i.e. lies below the deterministic frontier), the stochastic frontier output varies within the deterministic part of the frontier model. Thus, the deterministic production function lies in between the estimated stochastic frontier productions. In this model, it is possible to find the observed output above the deterministic frontier if the corresponding random errors are greater than the corresponding inefficiency effects ($v_i > u_i$).

The stochastic frontier model can be used with possible modifications and extension. Among the most frequently cited reasons for doing so are the general assumptions relating to the distribution of u_i (truncated or half-normal, etc.) and the consideration of panel data and *time-varying technical efficiencies*. Battese and Coelli (1995) proposed a stochastic frontier production for (unbalanced) panel data, which allows the firm's inefficiency effects to vary systematically over time. They also proposed computationally the easiest way of working with the time-varying inefficiency model. In the time-varying model, inefficiency coefficients are defined as an exponential function of time. Assuming TE as time-varying created a need to explain the factors which induce changes in the inefficiency effects. Policy reforms (such as market liberalization, credit institutions) and possible structural change in a sector are among the prominent causes of substantial changes in production efficiency over time.

Figure 4.3: Stochastic Frontier Production Function

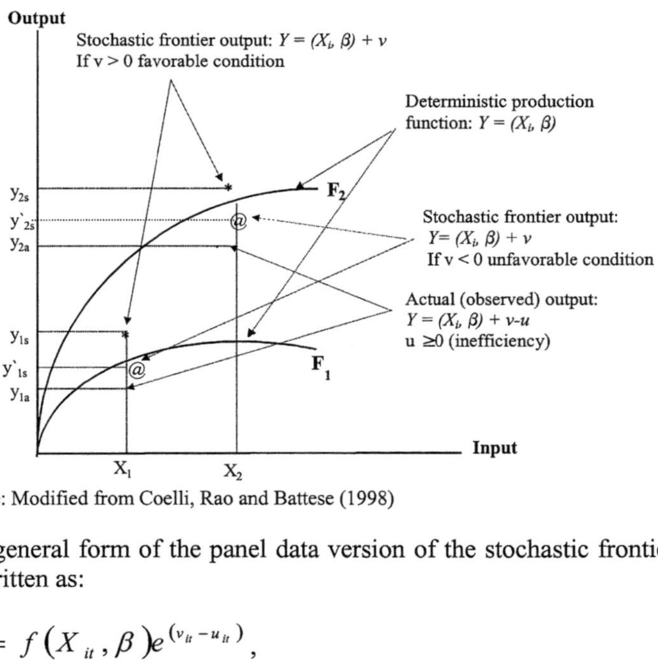

Source: Modified from Coelli, Rao and Battese (1998)

The general form of the panel data version of the stochastic frontier model can be written as:

$$Y_{it} = f(X_{it}, \beta)e^{(v_{it} - u_{it})},$$

or, in a simplified log linear form, it can be shown as:

$$\ln Y_{it} = \ln f(X_{it}, \beta) + v_{it} - u_{it}; \tag{4.1}$$

Where i denotes the i-th farmer or region, t denotes time; $f(X_{it}, \beta)e^{vit}$ is the stochastic frontier, where e^{ui} lies between zero and one; $f(X_{it}, \beta)$ is the maximum potential output for a particular input vector X_i; and β is the parameter vector of the production function which reflects the conditions that represent the production response of adopting "best practice" techniques. The variable Y_{it} is actual output for the i-th farmer in t-th period and X_{it} is a non-stochastic $(1 \times K)$ vector whose values are functions of inputs and other explanatory variables; v_{it} is a stochastic variable "white noise" representing measurement errors and other uncontrolled random shocks in production (such as the effects of weather changes, strikes, luck, etc.). It constitutes a two-sided error $(-\infty < v < \infty)$ and is assumed to satisfy the classical assumption: independently and identically distributed (i.i.d.) random error that has normal distribution with a mean of zero and unknown constant variance:

$v_i \sim N(0, \sigma_v^2)$, and $E(u_{it}, v_{it}) = 0;$

The variable u_{it} represents a systematic deviation from the frontier because of an unobservable random variable associated with technical inefficiency of the farmer in production. That is, for a given level of technology and inputs, the observed output falls short of its potential output. It is a one-sided (non-negative) error term ($u_i \geq 0$), implying that all observations lie beneath or on the stochastic production frontier. The u_{it} is assumed to be an i.i.d. truncated or half-normal random variable with a mean of zero and constant variance σ_u^2. For the given periods $t \neq t_1$, it is also assumed that $E(u_{it}, u_{it1}) = 0$ for all i.

Different specifications for u_{it} are proposed by Battese and Coelli (1992; 1995) particularly for panel data. In time-varying inefficiency effects of stochastic frontier modeling, u_{it} is defined as follows:

$$u_{it} = \{\exp[-\eta(t-T)]\}u_i; \text{ I} = 1,2,3,\ldots \text{N}; \qquad (4.2)$$

where the subscripts i and t represent the i-th firm and the t-th year of observation, respectively; η is an unknown parameter to be estimated, and u_{it} is an i.i.d. non-negative random variable obtained by truncated (at zero) or half-normal distribution with an unknown mean μ and constant variance, σ^2.

This model specifies that the technical inefficiency effects for the sample farmers in earlier periods of the panel are a deterministic exponential function of the inefficiency effects for the corresponding farmers in the last period of the panel (i.e., $u_{iT} = u_i$, given that the data for the i-th farmer are available in period,

T (Battese and Broca (1997). Given the assumptions of the stochastic frontier model, inferences about the parameters of the model can be based on maximum likelihood (ML) estimations because the standard regularity holds (Battese, 1992).[11] The ML requires numerical maximization of the likelihood function.

Following the procedure of ML estimation of the frontier production function, parameters such as γ and η deserve particular interest. For a balanced panel, as is the case in this study, the parameters μ and σ^2 define the statistical properties of technical inefficiency in relation to the last period of observation. The parameter γ can be expressed as $\gamma = \sigma_u^2 / (\sigma_u^2 + \sigma_v^2)$, which lies in the range of 0 to 1. Its maximum likelihood can be estimated using an iterative maximization algorithm (Coelli, 1996; Coelli, Rao and Battese, 1998). Accordingly, $\gamma = 0$ indicates that the deviations from the frontier are entirely due to statistical noise, while a value of $\gamma = 1$ would indicate that all the deviations are due to technical inefficiency. Given the specification of this model, there is a particular interest in testing the null hypothesis that the technical inefficiency effects are not present in the model. This is expressed by Ho: $\gamma = 0$.

The exponential parametric form of the technical inefficiency effects associated with time implies that if η is positive, then TE must be increasing at a decreasing rate; if η is negative, then it shows that the individual firm's TE is decreasing; and if $\eta = 0$, then TE remains constant over the given period. Therefore, a null hypothesis that the technical inefficiency effects are time-invariant is defined by Ho: $\eta = 0$; and that they have half-normal distribution are defined by and Ho: $\mu = 0$, respectively. These null hypotheses are tested using the loglikelihood ratios test.[12]

We suppose that the maximum capacity to produce is measured from observed input and output, based on the best practice performance of all the firms in the sample. Only when a firm uses a particular input with full efficiency does it receive the highest possible response from that input. A firm may or may not use all the inputs with full efficiency. TE of a firm *i* at time *t*, then, is defined

---

[11] There are also other estimation techniques (see Greene, 1993 for details). However, Coelli (1995) suggests that the ML method is the most applicable to estimation of stochastic frontier specifications. A detail for derivation of the logarithmic likelihood function for sample observations is documented in Battese and Coelli (1992).

[12] The generalized loglikelihood ratios statistic, λ, is defined by $\lambda = -2\ln (LHo)/L (H_1)$, where Ho and H_1 are the null and alternative hypotheses involved. If the null hypothesis, Ho, is true, then λ is asymptotically distributed as a chi-square (or mixed chi-square) random variable. If the null hypothesis involves $\gamma=0$, then λ has mixed chi-square distribution (see Coelli, 1995; 1996).

as the ratio of observed output to the corresponding frontier output, conditioned by the level of inputs used by that firm. This can be expressed as:

$$TE_{it} = \frac{Y_{it}}{Y_{it}^*} = \frac{f(X_{it}, \beta)e^{(v_{it} - u_{it})}}{f(X_{it}, \beta)e^{v_{it}}} = e^{-u_{it}} \tag{4.3}$$

Where Y_{it} is the actual observed output and Y_{it}^* is the maximum feasible output given by the stochastic frontier function. The maximum output stems from two sources: from the efficient use of inputs, and from firm-specific intrinsic characteristics including economies of scale. The latter is not explicitly included in the model but may produce a combined lump sum to the TE. The TE_{it}, therefore, is the expectation of the exponentiated technical inefficiencies, conditional on the errors. Since u_{it} is a non-negative random variable, the TE_{it} lies between zero and one, where one is the highest score indicating that the firm is technically perfectly efficient. Technical inefficiency is then given by: $TI_{it} = 1-(e^{-uit})$. Since u_i is unobservable, the technical efficiencies are predicted using the conditional expectation of $exp(-u_{it})$. Based on the conditional distribution of u_{it}, and given the distribution v_{it}-u_{it}, the efficiency of a specific firm at given time can be also given by:

$$TE = E\left\{\exp\left(\frac{u_{it}}{\varepsilon_{it}}\right)\right\} = \exp\left[\left(\frac{\sigma_u \sigma_v}{\sigma}\right)\left(\frac{f\left(\varepsilon_{it} \lambda/\sigma\right)}{1-F\left(\varepsilon_{it} \lambda/\sigma\right)} - \frac{\varepsilon_{it}\lambda}{\sigma}\right)\right] \tag{4.4}$$

Where $\epsilon_{it} = v_{it}$-u_{it}, and $\sigma = (\sigma^2_{u} + \sigma^2_{v})^{1/2}$ is standard error of $\epsilon_{it,}$; $\lambda = \sigma_u/\sigma_v$, it indicates the relative contribution of u and v in the determination of ϵ_{it}. The f (.) and F (.) are values of the standard normal density function and standard normal distribution function respectively evaluated at the $\epsilon_{it}\lambda/\sigma$. Along with the parameters of the function, the program also estimates $\gamma = \sigma_u^2/\sigma^2$, which is bounded between zero and one. The γ parameter is the ratio of the one-sided error term (inefficiency) to the total error. In other words, it measures the proportion of the total variation in observed output that is attributed to technical inefficiency.

4.2.2.2 Empirical Application of the Frontier Model in Previous Studies

Although the frontier production function is most often used for efficiency analysis, its application for analysis of policy impacts assessments is very recent and the literature is limited. Nevertheless, frontier models with the systematically varying technical inefficiency model specified above have been popularly used in policy analysis.

Fan (1991) applied this model using panel data from 29 provinces for the years 1965 to 1998 to estimate the effect of technological change and institutional reform on production growth in Chinese agriculture. He found a total output growth rate of 5.04%, of which 57.70% was attributable to increased use of total input and the remaining 42.3% to growth in TFP. About 63% of the TFP change was explained by institutional change (efficiency improvements) and about 37% by technological change. Fan then concluded that increasing use of traditional inputs, particularly chemical fertilizer, contributed significantly to the growth of the sector. The role of institutional changes in efficiency improvements and their subsequent impact on output growth was highly appreciated. He therefore proposed a new strategy focusing on regional specialization based on comparative advantages, and advocated discarding the policy of self-sufficiency at national and local levels. The implication was that to exploit the benefits deriving from efficiency improvements, crops should be grown where the soil and climate provide the most favorable conditions.

Similarly, Voigt and Uvarovsky (2001) used the frontier approach to evaluate the impact of transition on the development of productivity of Russia's agricultural sector in the transition period (1993-1998). They found that, assuming the natural conditions remained constant, the reliability of the political program, market reforms and restructuring of enterprises increased TE by an average of 7%, which surpassed the general trend of the nation. Moreover, TE increased from 1993 onwards. On the other hand, technological change was found to be substantially negative. Thus, TFP decreased by at least 20% in the study period, which ultimately led to a deterioration of Russia's agriculture. Therefore, it was suggested that there was an urgent need to strengthen incentives for investment in machinery and equipment.

The frontier production function model was also used by Salim and Kalirajan (1999) to analyze the relative importance of sources of growth in the food processing industries of Bangladesh before and after policy reform. The result showed that capacity realization (efficiency improvement) of firms was not substantial; it increased by only 7%, from 51% in 1981 to 58% in 1991, with considerable variation among firms. In a number of firms, efficiency had in fact deteriorated. The implication is that without further increasing input, there is still potential to increase output by improving the way in which it is applied. Although some firms experienced technological regress, it still accounted for the majority of TFP growth. Growth in TFP contributed 0.127% to 3.027 % of the total output growth per annum. On average, input growth continued to contribute the largest share (2.5%) of total output growth in the food processing industries in Bangladesh.

Battese and Broca (1997) employed a time-varying stochastic efficiency model for wheat farmers in Pakistan. Their results indicate that technical inefficiency decreased over time, but no technological change was observed. Curtiss (2002) intensively used the same model to analyze efficiency and structural change in the Czech Republic's crop production. He disclosed that the economic environment, particularly the functioning of the market and the uncertainties involved in it, has a greater influence on efficiency than internal firm-specific factors. Furthermore, he identified that production lines with the most sensitive and risk-averse organizational form showed the smallest productivity increase.

Many of the previous studies find different results. This implies that policy reform does not necessarily result in positive changes in firms' production behavior and production growth in an economy.

Box 4.1: Alternative Approaches to Assess Impacts of Interventions

Most conventional approaches to evaluate the effect of policy changes on the performance of an economy can be carried out alone or as a combination of the following approaches:

i) *Representing policy change with a variable*: - Quantifying the extent of policy changes induced or simulating with proxy variables and then measuring their partial effects on the expected resultants would have been the easiest and probably the best approach. However, institutions as cost carriers are not considered as production factors, and are difficult to quantify to capture their effect on production and development. Neither is possible to represent the likely change in the structure of production after the reform by dummy variables, as household data series for the years before the reform were not available. Such an approach is possible if aggregated national data is used, but may give only crude information.

ii) The *"with-and-without"* approach: - This approach is used to compare, for example, a region or a district of a country which assumed policy or strategic change versus those with no change using certain selected criteria such as changes in economic indicators. Since the reform has encompassed virtually the whole country, this approach also cannot be used. It might be possible (with strong reservations) to undertake a country comparison, but almost every country has either already adopted economic reform or is in the process of undertaking the reform. One can attempt a *dummy variable regression*, in which case it becomes similar to the above approach. Allowing one of the years as a base year, in panel data (like in our case) dummy variables for each cross-sectional observation should be included, along with explanatory variables. However, in most cases, this results in too many explanatory variables to explicitly carry out the regression (Wooldridge, 2003). Thus, the dummy variable method is not very practical for panel data sets with many cross-sectional observations.

iii) The *"before-and-after"* approach: - In this approach, the impact of a development strategy or policy change is linked to a deviation in certain measurement criteria from their initial levels before the reform took effect. Although data for the period before the reform were not available, a normal year before the reform can be selected and used as the base year for comparison if there are changes in the years thereafter. Nevertheless, this still fails to account for the possibility of changes (in production) that would occur even without the reform (Gittinger 1982).

Source: Own reading from different sources

4.3 Research Design and Data Sources

As rural development is the central focus of the government, this study also concentrates primarily on the agricultural sector. Farm households should be the first beneficiaries of good policies and development strategies, or otherwise they are victims of it. Aggregated micro effects constitute sectoral and national economic changes. Consequently, this research dealt mainly with analyzing changes in major farm production of rural households.

A rural database already developed by the Department of Economics of Addis Ababa University in collaboration with the Center for the Study of African Economies (CSAE) and Oxford University for the years 1994, 1997, and 1999 was used. In addition, one-year data for the year 2001 were collected for the same households to complement the data of the AAU. Thus, a four-year panel of data was used to analyze the performance of agricultural production and impacts at household level.[13] In the early years following the withdrawal of the socialist government in 1991, the primary activities of the EPRDF government were to stabilize the political mess and to draft development strategies. A series of policy and institutional reforms have been undertaken since 1992. Although such activities have been an ongoing process, the rural-based development strategy came into effect in late 1994. The year 1994 was therefore selected as the base year for two main reasons. First, it was the year in which the government commenced its rural-based development strategies, the effect of which (i.e., changes in the production of rural households) might start to become apparent in the succeeding years. Second, it was a normal year, one in which the political and economic situation was relatively stable. We presumed that after this important period, farming would be in a state of disequilibrium (transition); there would be dynamic positive changes in the components of output growth.

On the whole, the period 1994-2001 can be characterized as a period in which the nation experienced relatively normal economic indicators (e.g., prices), and normal climatic conditions (rainfall) and was a peak period for technology transfer activities. Therefore, the results of the empirical analysis would indicate that the estimated progresses (or regresses) are potential changes under normal economic and climatic conditions, and when institutional support in terms of input and technology transfer was relatively at its highest.

---

[13] For this purpose, the panel data was found to be richer in information than time-series data (as the latter is too small in size since the reform and is also highly aggregated).

4.3.1 Selection of the Study Areas

Although policy and technological innovation play a significant role in the growth of agriculture, its diverse production potentials are largely determined by the natural biological and physical characteristics of the environment. Besides development policies and strategies, differences in physical environment such as amount of land owned, soil quality, rainfall, etc., could have considerable influence on production technologies and efficiency of production. Accordingly, farmers' production systems and preferences change as a result of the cumulative effects of the aforementioned factors. In such cases, both highly aggregated and small location-specific studies related to rural development could equally lead to biased conclusions or provide inadequate or only partial information. In Ethiopia every rural people, in all corners of the country, tries to grow crops (except the nomads) and to rear livestock. However, only parts of the midlands and highlands, characterized by a moist, sub-humid or humid climate, have reasonable potential for agricultural. These areas embody the largest source of agricultural production, and represent over 95% of total farm production at national level. More than 85% of the human population and over 70% of livestock are also concentrated in these zones.

Many authors have described the agro-ecological zones of the country differently at different times. Two important and reasonable classifications are referred to here. Amare (1978) classifies the agricultural system of Ethiopia into three zones based on national resource endowment, environment and economic grounds. These include high-potential cereal/livestock zone, low-potential cereal/livestock zone and a high-potential horticultural/livestock zone. A similar but more detailed classification of the land use system is given by ICRAF (1990) based on biophysical and socio-economic factors. ICRAF also classifies the land use system of the country into three major agro-ecological zones: high-potential perennial crop, high-potential cereal and low-potential cereal zones. The high-potential perennial zone covers some parts of the south-west of the country. It represents the hilly and intensive cultivation of enset *(false banana),* and major coffee growing areas. The high-potential cereal zone covers vast areas of Arsi, Bale, Wallaga, Showa, and Gojam. It is further divided into mixed cereal-livestock (that extends from the altitude of 1500 to 2500 meters above sea level) and barley-livestock-based systems (above 2500 meters). The low-potential cereal zone is located predominantly in much parts of Tigray, North Gondor, Wollo and the arid eastern lands of the country.

As the country's main objective is first and foremost to satisfy the demand for food, we were interested primarily in studying cereal-based production systems. In the first stage, therefore, the high-potential cereal zones and the low-potential cereal zones were selected for study because of their relative importance in agriculture. As may be noticed, neither the Amare's nor the ICRAF classifications includes any medium-potential zone. However, upon closer observation of the study sites, and based on the information we obtained from descriptions of location-specific socio-economic parameters, we designated the high potential barley-livestock ecological zone as a medium cereal-livestock zone. In this way we selected three different agro-ecological zones with high, medium and low potential.

Taking into account fair representation, proximity to the main roads and experience in the use of modern farm inputs, three administrative districts, each representing different agroecological zones (agricultural potential) was chosen. These include *Shashamane, Debrebrehan Zuria* and *Ankober-Dinqi*. Shashamane district represents the high-potential zone; Debrebrehan Zuria district represents the medium-potential zone and Ankober-Dinqi the low-potential zone. Shashamane is located in the Rift Valley of Ethiopia in the south-east of Oromia Regional State on the road to Awasa, at about 230km from the capital, Addis Ababa. Both Debrebrehan Zuria and Ankober-Dinqi are located in the center of the high plateau of Ethiopia in the Amahara Regional State. These districts are represented by Debrebrehan town, which is located at about 120km to the north of Addis Ababa on the road to Wollo or Tigray region. In terms of transport, communication and access to market infrastructure, Shashamane district is in the most advantageous position.

4.3.2 Physical Characteristics of the Study Areas

Table 4.1 summarizes the physical characteristics of the study areas. Shashamane is situated in plane topography with a mid-altitude of about 2000 meters above sea level. It is endowed with a favorable temperature that ranges from 12°C to 30°C. Compared to other locations, the soil in this area is regarded as fertile and suitable for agriculture. According to ICRAF (1990), croplands are covered with a mixture of different types of soils. Shashamane receives an average annual rainfall of about 800mm, which is characterized by bimodal rainy seasons, *meher* and *belg*. The *meher* season refers to the main rainy season corresponding to the main cropping period, in which the area receives over 70% to 80% of the total annual rainfall. This period starts in June and lasts until mid-September. *Belg* is a short rainy season in March and April. However, the *belg* season is generally risky and highly unreliable because of high variability, delay and even absence of rainfall. The other two study areas also have a similar bimodal pattern of rainfall. Shashamane is suitable for a variety of crops, the major ones being wheat, barely, tef, maize, potato and onion.

Debrebrehan Zuria is a typical highland sub-humid area. Its altitude ranges from 2400 to 2800 meters above sea level. It is among the coldest areas in the country, and frost is a common occurrence. The annual minimum average temperature is 10°C and the maximum average is 15°C. The topography of Debrebrehan Zuria is characterized as flat-to-undulating and hilly types and is vulnerable to run-off. Heavy-textured black clay soils (vertisols) dominate in the flat areas, associated with luvisols in areas with moderately undulating topography. Hence, drainage and lack of fertility are critical problems in this area. There is a unique traditional cultural practice of soil burning (*guie*) by farmers in this area, which slightly improves the soil's texture and fertility (ICRAF, 1990). The average annual rainfall is about 1100 mm. Despite its risky nature, *belg* is very important for crop production in the highland areas, as frost and drainage problem are serious during the *meher* season. The main crops grown in Debrebrehan Zuria include: barely, wheat and different pulse crops integrated with relatively the highest population of livestock.

Ankober-Dinqi is a mountainous district with rugged terrain. The altitude ranges from as low as 1500 to 2000 meters above sea level. Accordingly, rainfall intensity varies from an annual average of 900 mm at the lowest altitude to more than 1400 mm at the highest altitude. The mean temperature also ranges from a minimum of 15°C at the highest peak to a maximum of 25°C at the lowest altitude. Correspondingly, a variety of soil types is present and hence different kinds of crops are grown (Senait, 2002). In this district, our study concentrated particularly on households located at lower altitudes, as these are more representative of the zone of low agricultural potential.

Table 4.1: Topography and Agro-Climatic Description of the Study Areas

Characteristics	Shashamane	Debrebrehan	Ankober-Dinqi
Altitude (meters)	2000	2400-2800	1500-2000
Average annual rainfall (mm)	800	1100	900-1400
Average daytime temperature	25.0°C	12.6°C	15-25°C
Soils	Mixture of vitric, mollic & andosols	Predominantly vertisol	Mixture of many types
Topography	Plane	Plane & rugged	Rugged
Major crops	Wheat, barley, tef, maize & vegetables	Barley, wheat, and beans	Sorghum, tef, maize, pulses

Source: Different Series of Ethiopian Village Studies (AAU)

4.3.3 Selection of Production Lines

Studies have ascertained that individual and aggregated outputs respond differently to policy changes (Binswanger, 1989; Schiff, 1997), with the former being more responsive. In order not to underestimate the role of policy changes in Ethiopia, therefore, this study treated each crop separately. Cereals have certainly become a priority area of research because they dominantly contribute to the agricultural GDP (60%) and domestic food production (80%). In Ethiopia, cereals account for 82% of the cropland and 87% of the total annual crop output.

According to CSA (2003), "tef" *(Eragrostis tef)*, maize, wheat, sorghum and barley are the major cereals food crops in order of importance. In this study, the first three crops were given particular attention, as they are particularly characterized by different technologies as a mirror image of diverse factor specificity, intensity and organization. Thus production lines tef, wheat and maize were selected for analysis. Two key criteria were used to select the production lines for the empirical analysis: first, their relative importance in agriculture and as a source of food; and second, expected changes in the technologies used for their production because of the particular emphasis placed on them by research organizations and extension services. Tef, wheat and maize together constitute the major food crops, sharing on average about 78% of total crop area, and 71% of the total output of cereal crops. Tef is the most widely grown and most input-demanding crop (in terms of labor, land and fertilizer). It is also known to be the most important staple food, especially in towns and in the northern part of the country in general. A usable improved variety has not, however, been developed for this crop. On the other hand, maize and wheat are probably the only two important crops for which alternative production technologies have been developed and widely diffused to rural farmers, particularly since 1994. On the whole, these three crops are economically most important and have received top policy attention in attempts to improve food security and agricultural development.

Generally, highland crops such as barley and wheat are agro-ecologically unsuitable in Ankober-Dinqi, but are most important in Debrebrehan. On the other hand, sorghum, maize and tef are important in Ankober-Dinqi, but not so in Debrebrehan. Shashamane is suitable for almost all kinds of crops. Farmers in Shashamane district therefore do have better possibilities for choosing which crop to grow based on the socio-economic circumstances they face. The importance of the three major cereals focused on by the research (tef, wheat and maize), varies in different locations. Thus, depending on their relative importance in the selected districts, we considered data relating to tef and maize from the districts of Shashamane and Ankober-Dinqi, and data on wheat from Shashamane and Debrebrehan.

4.3.4 Sampling Design and Method of Data Collection

In Ethiopia in general and in the study areas in particular, farmers in a district are organized into Peasant Associations for political and economic objectives. The number of farm families in each Association is neither fixed nor uniform, but a simple collection of farmers within geographically similar areas. Usually, each Association comprises over 500 farm families. For this micro-level farm household survey, a two-stage random sampling technique was applied. In the first stage, Associations were selected by a simple random sampling method within the selected districts. In the second stage, farm families represented by household heads were selected randomly for interview. The number of Associations selected from each district was not equal, and neither was the number of families selected from each Association. It depended on the size and homogeneity assumed in each case. The higher the population and the lower the homogeneity in the farming system, the larger the sample size, and vice versa.

A standardized questionnaire was then used to conduct the interview. It was first tested for modifications before the surveys were actually carried out. The questionnaire was designed in such a way as to include important information such as household demographic characteristics, resource endowment, physical quantities of inputs used and grain outputs produced, institutional supports such as access to credit and transfer of improved technologies, perceptions about changes and level of output and factor prices, and participation in off-farm activities. Data were collected by trained enumerators by means of interviewing the household heads of the randomly selected farm families. In addition, extensive discussions on farming activities as a whole and problems encountered were held with groups of farmers and development agents in order to substantiate the information provided in the questionnaire. Panel data covering four years, 1994, 1997, 1999 and 2001 were collected from a total of about 750 sample farm households.

4.4 Statistical Description of Changes in Resource Endowment and Use

Land and livestock constitute the major component of farm resources in the mixed farming system. Changes in natural factors such as population pressure, climatic conditions and government intervention can induce changes in the relative importance of the size of these resources owned by individual households through their effects on product and factor markets. In addition, institutional changes can provoke farmers to use new and modern inputs, and cropping systems may also change accordingly. It is often not easy to filter out the effect of the policy reform on farmers' resource endowment, and so in this section we address only the general direction of changes following the reform. Nevertheless, changes in resource use can largely be ascribed to the effect of policy changes as they influence incentives and the availability of modern farm inputs such as chemical fertilizer and genetically improved seeds.

4.4.1 Livestock and Draft Power

Livestock constitutes the main share of capital in a traditional mixed farming system. In addition, livestock is the main source of farm income, and the amount of livestock belonging to a farm directly determines the level of production and productivity of crop production as a source of labor power and of manure, and indirectly by increasing the probability of obtaining loans for the purchase of improved technologies. On the one hand, farmers mostly purchase livestock from the savings they obtain in years of good harvest. On the other, they sell off livestock during years of bad harvest to secure their food supply. As financial institutions are rudimentary, changes in livestock owned are the best measure of capital development in rural mixed farming communities, *ceteris paribus*. Table 4.2 shows the mean and range of total livestock units (LU)[14] and oxen/bulls owned by sample farmers over the study period.

Table 4.2: Descriptive Statistics of Trends in Livestock Units and Oxen Owned per Farm Household

Descriptions		Year			
		1994 (n=350)	1997 (n=335)	1999 (n=300)	2001* (n=210)
Total livestock units owned	Minimum	0.00	0.01	0.00	1.00
	Maximum	28.27	27.67	23.22	21.25
	Mean	6.91	6.04	5.76	5.23
	Std. deviation	5.77	4.70	4.80	4.39
Number of oxen/bulls owned	Minimum	0.00	0.00	0.00	0.00
	Maximum	9.00	6.00	7.00	8.00
	Mean	1.52	1.58	1.80	1.93
	Std. deviation	1.60	1.26	1.39	1.55

Source: Own computation from the survey data

Note: Sample size in 2001 was much less than in the preceding years because it was preferred to make more intensive and better quality. The information provided was consistent.

In the years under consideration, while the mean LU per household steadily decreased, the mean number of oxen/bulls gradually increased; both were significant at the 5% level (Table 4.3). Compared to the base year, 1994, a significant decline in LU was noted in 1999, but the increase in the number of oxen/bulls was significant in 2001 (Table A4.1 in the Annex). This implies that sample farmers took sufficient time to adjust their resources. The opposite direction of changes in the development of LU and oxen could be due to the need to keep fewer livestock in general, possibly because of the worsening problem of shortage of grazing land. In other words, farmers need (or are forced) to keep the most essential and most productive livestock classes such as oxen and milking cows. Keeping large numbers of livestock for social esteem and/or for the purpose of transferring savings is on the point of losing its place because of the ever-diminishing land holdings due to population pressure.

---

[14] Computation of Livestock Unit (LU) is adopted from the conversion factor developed by Herman Strock et al. (1991).

Further comparative analysis across the locations of the study shows that there was a significant difference in total LU and oxen owned in the different locations over the study period. Bonferroni multiple comparison analysis indicates that sample farmers in Debrebrehan had the largest livestock and oxen population, followed by farmers in Shashamane. Ankober-Dinqi was the least endowed in terms of both LU and oxen/bulls (Table A4.2 in the Annex). However, both LU and oxen increased in Ankober-Dinqi district. They increased from a mean of 2.35 LU and 0.56 oxen/bulls in 1994 to a mean of 2.90 LU and 1.40 oxen/bulls in 2001. In the other locations, however, there was a slight decline in the number of LU but an increase in oxen.

Table 4.3: ANOVA for Changes in Livestock and Oxen Capital (1994-2001)

Dependent variables	Source of variation	Sum of squares	df	Mean square	F
Total livestock units owned	ESS	666.79	3	222.264	8.74**
	RSS	28856.19	1135	25.424	
	Total	29522.98	1138	-	
Number of oxen and bulls owned	ESS	100.945	3	33.648	16.23**
	RSS	2448.216	1181	2.073	
	Total	2549.161	1184	-	

Source: Own computation
Note: **The mean difference is significant at the 0.05 level

Table 4.4: Changes in the Relative Proportion of Sample Households Owning Oxen/Bulls over Time

Number of oxen and bulls owned	Year			
	1994 N = 372	1997 N = 357	1999 N = 338	2001 N = 218
Zero**	34.13	26.03	20.43	15.84
One	25.81	20.44	20.68	27.44
Two**	15.85	31.36	35.49	31.60
More than two	24.19	22.12	23.40	25.18

Source: Own computation
Note: ** Change over time is statistically significant at 5% level.

On average, 45% of the sample farmers in Ankober-Dinqi, 32% in Shashamane and 13% in Debrebrehan were without oxen during the years analyzed. Lack of oxen-power was severe in Ankober-Dinqi and Shashamane. In all cases, the percentage of families owning zero oxen fell much faster in Ankober-Dinqi than in the other regions. For instance, in the base period, about 60% of sample farmers in Ankober-Dinqi, 19% in Debrebrehan and about 39% in Shashamane were without oxen. This number fell to 31%, 8% and 24% respectively in 1999/2001. Consequently, the proportion of farmers having a pair of oxen gradually increased over the study period (Table 4.4; and Table A4.3 in the Annex).

4.4.2 Land Resources

Land is the most decisive resource in agriculture. The amount of farmland owned and its qualities (e.g., fertility and suitability for the desired farming system) have considerable impact on the livelihood of traditional farmers. Given the technology of farming in Ethiopia, land is generally said to be scarce, and the scarcity is very severe in areas of higher agricultural potential because of increasing population pressure. Table 4.5 indicates changes in the mean area of land owned across the districts in two distinct periods. Although there appeared to be a general declining trend in the mean size of land owned, no statistically significant change was observed over the period under consideration. The mean land size owned per sample household is a little more than three hectares in Debrebrehan, and less than 1.5 hectares in Shashamane and Ankober-Dinqi. As indicated in Table 4.6, a Bonferroni multiple comparison test shows that, relatively, Debrebrehan farmers owned and cultivated the largest amount land per household, but the difference between Ankober-Dinqi and Shashamane was not substantial. Analysis of variance also showed similar results at a 5% level of significance (not shown here).

Table 4.5: Changes in Land Owned and Land Used

Land use	Ankober-Dinqi		Debrebrehan		Shashamane	
	1994	2001	1994	2001	1994	2001
Average total land owned (ha)	1.43	1.27	3.22	3.12	1.25	1.11
Cropping land (%)	90	91	71	76	98	97
Fallow land and/or grazing (%)	30	19	82	70	13	8

Source: Computed from survey reports

It is apparent that the proportion of land allotted for fallow/grazing was generally diminishing. Except in Debrebrehan, there is little difference in the amount of land cultivated and total land owned. In other words, farmers cultivated almost all the land at their disposal, abandoning fallowing and reducing grazing land. In Shashamane particularly, nearly all the land owned was allocated for crop production (97% in 2001 and 98% in 2001). The proportion of land owned that was assigned for fallow and/or grazing was highest in Debrebrehan (about 29% in 1994, although the share fell to 24% in 2001), where the highest proportion of sample farmers (over 70%) engaged in this practice. Land was most intensively used for cropping by the sample farmers in Shashamane and Ankober-Dinqi, where considerably fewer sample farms were found to practice fallowing and/or leaving land for grazing. About 40% of the sample in the latter two locations cultivated over 90% of total land they owned, and about 33% cultivated all of their land.

Differences in land and livestock (oxen) endowment between the different locations clearly imply a difference in agricultural potential, with all other factors held constant. In this sub-section, we can conclude that the three locations of the study were substantially different in terms of livestock and land endowment over the study period, which could make a vital difference as regards their agricultural production potential and ultimately their livelihoods.

Table 4.6: Multiple Comparisons of Total Land Owned and Cultivated Across the Locations of the Study

Dependent variable	(I) Location	(J) Location	Mean difference (I-J) Bonferroni
Total land owned (ha):	Ankober	Debrebrehan	-1.980*
		Shashamane	0.105
	Debrebrehan	Ankober	1.978*
		Shashamane	2.083*
Total land cultivated (ha):	Ankober	Debrebrehan	-1.008*
		Shashamane	0.069
	Debrebrehan	Ankober	1.008*
		Shashamane	1.077*

Source: Own computation
* The mean difference is significant at the .05 level.

4.4.3 Factor Market and Modern Input Use

4.4.3.1 Labor and Land Transactions

Land and labor transactions in Ethiopia are generally very rudimentary because market institutions and information facilities are not yet well developed. In the pre-reform period (during Socialism), renting land and hiring labor for a wage were prohibited by law. Instead, only communal or cooperative work was encouraged. This legal and political system was in force for such a long time in the rural sector that it may still influence the mentality of farmers, making them reluctant to get involved in land and labor transactions even after the market reform (56% of sample farmers reported this in the 2001 survey report).

Table 4.7: Proportion of Households (%) Involved in Hiring Labor (in/out)

Locations	Year			
	1994	1997	1999	2001
Ankober-Dinqi	3.40	14.00	10.60	12.5
Debrebrehan	17.00	23.00	40.70	30.70
Shashamane	32.00	24.60	52.00	37.30
Total sample	18.04	21.51	36.53	35.32

Source: Computed from the survey reports

As may be seen in Table 4.8, participation of sample households in the labor market seems to improve slightly over the years considered. The proportion of sample households involved increased from 18% in 1994 to about 35% in 2001. In each location, there was a clear increase in the percentage of farm households either hiring labor in or out. The growth in labor transactions among the locations is quite different, being highest in Shashamane and lowest in Ankober-Dinqi. This could be because of differences in people's increasing awareness of external sources of income and/or increasing the intensity of labor use in the agricultural production process (in case of hiring in labor). Hired labor was used, to a greater extent, for sharecropping than on self-cultivated land, and more (75%) for harvesting and weeding activities. However, this growth suddenly decreased in the 2001, perhaps due to the decline in output prices in that year (discussed in the next chapter).

Table 4.8: Relative Importance of Households Involved in Renting Land (in/out) and Sharecropping (in/out) (in %)

Year	Land rent in/out				Land share in/out			
	Ankober-Dinqi	Debre-brehan	Shasha-mane	Total	Ankober-Dinqi	Debre-brehan	Shasha-mane	Total
1994	1.2	3.0	15.0	5.87	11.5	18.5	25.5	18.78
1997	3.5	9.3	13.0	8.97	15.0	28.5	26.0	24.58
1999	7.3	15.0	27.0	16.58	58.0	56.0	42.5	52.76
2001	9.0	18.0	25.3	22.52	40.4	36.4	31.0	32.35

Source: Computed from the survey reports

As already mentioned in Chapters 1 and 3, farmers in Ethiopia are generally not allowed to sell or buy land because of land policy restrictions. Thus, renting and sharecropping were the only alternatives for obtaining more land during the period under consideration. Neither was it possible to expand cropland by clearing forest or bush lands since these had already vanished. In the event that a household failed to cultivate its land, most probably due to lack of oxen or other inputs, the household could make the land available for rental or sharecropping.

As may be observed in Table 4.7, the proportion of households involved in land rental and sharecropping transactions steadily increased over time. Households tended to be involved to a greater extent in sharecropping than in renting land. Both renting and sharecropping transactions were directly related to the agricultural potential of the locations, being most intense in Shashamane and least intense in Ankober-Dinqi.

4.4.3.2 Improved Seed and Fertilizer Use

The most familiar ways of intensifying agriculture are using chemical fertilizer, improved seeds and irrigation. As already discussed in Chapters 1 and 2, irrigation is little practiced in Ethiopia. Use of modern inputs is also very rudimentary. Nevertheless, there was an increasing trend in the number of farmers using commercial fertilizer and improved seed in the period referred to (see Table 4.9). Perhaps the most readily available improved seeds, both in the market and via extension agents, were for wheat and maize, and farmers also had greater awareness and experience of these. As indicated in the Table, the share of total sample farmers using improved wheat and maize seeds increased appreciably (by more than double) over the period analyzed. However, there were clear regional differences, with use of both improved wheat and maize being greatest in Shashamane and least in Ankober-Dinqi (see also Table A4.4 in the Annex). This may have been because Shashamane is potentially a more favorable area for agriculture, and because it has better access to transport networks and market centers.

On the other hand, improved seeds for other crops were at different stages of development in the research centers, and mostly at the stage of learning and/or in the process of diffusion. A great deal of work has still to be done in terms of the invention and innovation process for technological development of other crops as a whole.

Although, at national level, wheat, tef, maize and barley are the major crops receiving fertilizer in order of importance, the order of importance was completely different in our study areas. Table A4.4 in the Annex shows that sample farmers in the different locations were using fertilizer disproportionately. As was the case with the use of improved seed, chemical fertilizer was also most used by farmers in Shashamane followed by those in Debrebrehan and lastly in Ankober-Dinqi.

Three important points can be deduced from this table. First, the proportion of farmers using fertilizer and the rate of their utilization were markedly different, and this difference corresponded exactly to their order of agricultural potential. This implies that areas of high agro-ecological potential most probably adopt improved/modern practices because they are most profitable and are associated with fewer risks such as lack of rainfall, pest damage, etc. Second, across the period examined, except in 2001 (perhaps because of the fall in crop prices), the proportion of sample farmers using fertilizer increased markedly in all locations. This is an encouraging trend to alleviate the problem of food insecurity, especially if complemented with other basic factors like favorable market and weather conditions. In 2001, the number of farmers using fertilizer decreased relative to the preceding year, mainly owing to the fall in prices of agricultural outputs in that year. Third, depending on their objectives and ecological suitability, farmers give different levels of fertilizer concentration to different crops. For instance, while most farmers in Shashamane applied fertilizer to the greatest extent on wheat, this was the case for barley in Debrebrehan and tef in Ankober-Dinqi. Note that the sample farmers in Shashamane nevertheless used much more fertilizer on tef (approximately double) compared to farmers in Ankober-Dinqi.

Table 4.9: Percentage of Sample Farmers Using Fertilizer and Improved Seed

Crops	1994	1997	1999	2001
Chemical fertilizer:				
maize	33.2	39.5	59.0	74.0
wheat	34.0	53.5	62.5	64.8
tef	36.9	46.7	67.0	65.0
Improved seed:				
maize	6.5	11.0	17.7	25.5
wheat	2.3	8.2	15.6	19.5

Sources: Own computation

Adoption of improved technologies is a function of many interrelated socio-economic and natural factors. Among other things, the effectiveness of agricultural research and extension institutions in developing and diffusing adoptable technologies plays the greatest role. Yield and price stability are also among the major concerns behind risk-adverse behavior of peasant farmers. Thus, adoption of new technologies is about more than just profit maximization. The inter- and intra-regional differences in the use of modern inputs, improved seeds of wheat and maize and chemical fertilizer, could therefore be ascribed to differences in the households' socio-economic and natural endowments, which are not randomly distributed among them. Abebe and Mulat (2003) underlined that shortage in the supply of credit and high interest rates are the most critical constraints to use of improved technologies in Ethiopia. They described the credit market in Ethiopia as narrow, segmented and underdeveloped, and unable to mitigate the shortage of credit. As a result, few households are able to use modern inputs consistently, and of those farmers who do, a considerable number apply inputs below the recommended rate. This affects the efficiency and productivity of farmers. Consequently, the success of any policy intervention aimed at bringing about technological change in agriculture depends on the extent of diffusion and adoption of improved technologies. Improved technologies and modern inputs in general, must be used by a significant number of peasants in order to bring about meaningful changes in output growth.

4.4.4 Relative Changes in Land Allocation to Major Crops

On the whole, in line with the shrinking trend as regards size of cultivated land, the total land allocated to major cereals in particular was significantly reduced in each district compared to the base year (Table 4.10). Results of comparison of the study locations reveals that farmers in Shashamane used the lowest proportion of their cultivated land for cereal crop production. One major reason for this might be the increasing trend towards crop diversification, such as increasing cultivation of vegetables (onion, potatoes, cabbage, etc.) in Shashamane, and cultivation of oil crops in Ankober-Dinqi and Debrebrehan. Vegetables and oil crops are naturally produced mostly for sale.

Basically, the predominant crops in terms of area and number of households growing them historically in the farming system are sorghum in Ankober-Dinqi, barley in Debrebrehan, and wheat and maize in Shashamane. However, during the study period changes emerged in the proportion of land allocated and number of growers. We were therefore keen to gain a clearer understanding of changes in the trend of cropping patterns and estimate their extent using statistical tools. In the following sub-section, we will discuss in brief the statistical method applied and then move on to discuss the results for each location studied.

4.4.4.1 Statistical Method of Analysis

Tests for the statistical significance of changes in the proportion of land allocated to different crops and the number of sample growers were carried out according to methods based on the work of Berry and Lindgren (1996). Suppose n_1 and n_2 are independent sample sizes drawn from a population at a different time, and p_1 and p_2 are proportions (probabilities of success) of n_1 and n_2. $P_1 = y_1/n_1$, and $p_2 = y_2/n_2$, where y_1 and y_2 are the number of success (which in our case means area under cultivation with a particular crop, or number of households growing a particular crop in a specified year). The null hypothesis is that there is no difference between p_1 and p_2. The approach is specified as follows:

Ho: $p_1=p_2$, or equivalently,
$D = p_1 - p_2 = 0$, then it follows, (4.5)
Ho: E (D) = 0;

D is the difference in sample proportion in two periods that implies the difference in population proportion. In a large sample the distribution of D is approximately normal. By computing the mean and variance of D, it can be transformed into z-statistics for testing the null hypothesis, Ho. Because the samples are independent, the variance of D is the sum of the variance of p_1 and p_2.

$$S^2(D) = S^2(p_1) + S^2(p_2) = \frac{p_1(1-p_1)}{n_1} + \frac{p_2(1-p_2)}{n_2} \qquad (4.6)$$

Under the null hypothesis $P_1 = P_2$; and representing both by P, then the variance of P can be calculated as follows:

$$S^2(D) = \frac{p(1-p)}{n_1} + \frac{p(1-p)}{n_2} = p(1-p)(\frac{1}{n_1} + \frac{1}{n_2}) \qquad (4.7)$$

Where P is the pooled or weighted average the proportions of the two samples; its maximum likelihood is estimated easily as:

$$p = \frac{y_1 + y_2}{n_1 + n_2} = \frac{n_1 p_1 + n_2 p_2}{n_1 n_2}$$

From equations of 4.5, 4.6 and 4.7 a z-score can be calculated as:

$$Z = \frac{D - 0}{sd(D / H_o)} = \frac{p_1 - p_2}{\sqrt{p(1-p)(\frac{1}{n_1} + \frac{1}{n_2})}}. \qquad (4.8)$$

The test of a significance can be proceeded by comparing the observed difference ($D= p_1-p_2$) with the probability value under the standard normal cumulative distribution function to the right of the computed z in equation (4.8). This test compares the proportion of only two samples, and fails to show the trend of changes over a period. Thus, only the proportions of the base period and the last survey year were compared in this analysis. Indeed, changes between successive years were slight and not so informative. Nevertheless, since proportions of samples for each year are also included in the result, the direction of the changes is more easily discernible.

4.4.4.2 Results and Discussion

Because the research areas differ in terms of cropping system, changes in the cropping pattern are better described for each separately. Table 4.10 summarizes the significance of changes in the proportion of land allotted to major crops and in the proportion of sample farmers growing the crops in the study period.

Ankober-Dinqi: In Ankober-Dinqi, the proportion of maize and tef growers increased spectacularly, from 35% and 56% respectively in 1994 to 52% and 79% in 2001, (significant at 5% level). The mean proportion of land allocated for tef production increased from 21% in 1994 to 36% in 2001. The share of the maize area also changed in a similar direction but the change was not statistically significant. This resulted in a consequent reduction in the proportion of land area allocated to the dominant crop, sorghum, from 63% in 1994 to 48% in 2001. No appreciable changes were observed for other crops, however.

Debrebrehan: In Debrebrehan, wheat increased dramatically, taking over land and labor power from barley. The proportion of sample farmers growing wheat increased from 71% in 1994 to 87% in 2001, and its mean area increased from 18% to 26% (both significant at 5% level). Land share for beans and peas also increased significantly in the same period. Given the limited availability of land, the proportion of land allocated to barley (the dominant crop) decreased continuously and significantly. The land share for barley was 67% in the base period, falling to 43% in 1997 and rising again to 50% in 2001. However, almost all sample farmers continued to grow the crop throughout whole period.

Shashamane: In Shashamane too, changes in the cropping system are apparent. Farmers were gradually switching from production of barley, tef and sorghum to produce more wheat and maize. The proportion of barley and tef growers declined from 36% and 60% in 1994 to only 11% and 36% respectively in 2001 (significant at 5% level). Besides, the area used for minor crops fell radically from its share of 28% in 1994 to close to zero in the last year (not indicated in the table). Although there was no dramatic change in the proportion of growers, the share of land allotted to wheat and maize increased. The relative area of wheat increased from 30% in the base period to 42% in 2001. Similarly, the area used for maize rose from a mere 19% in the base period to 30% in 2001, (significant at 10% level). In 2001, wheat and maize together accounted for 72% of the total land cultivated for major crops in Shashamane.

Table 4.10: Trends in Cropping Patterns of Major Crops: Proportion of Land Allocation and Sample Farmers Growing the Crops

Crops	Ankober-Dinqi				Debrebrehan				Shashamane			
	1994 (i)	1997	2001 (j)	D[16] (j-i)	1994 (i)	1997	2001 (j)	D (j-i)	1994 (i)	1997	2001 (j)	D (j-i)
Barley	uni	uni	uni	uni	67 (100)	43 (94)	50 (99)	-17* (-1)	7 (36)	6 (26)	3 (11)	-4 (-25)*
Wheat	uni	uni	uni	uni	18 (71)	26 (81)	26 (87)	8* (16)*	30 (74)	41 (80)	42 (76)	12 (2)
Maize	8 (35)	16 (46)	15 (52)	7 (19)*	uni	uni	uni	uni	19 (72)	29 (86)	30 (84)	11 (12)*
Sorghum	65 (100)	46 (91)	47 (86)	-18* (-11)	uni	uni	uni	uni	3 (21)	5 (43)	8 (38)	5 (17)
Tef	21 (56)	30 (67)	36 (79)	15* (23)*	uni	uni	uni	uni	13 (60)	11 (24)	13 (36)	0 (-24)*
Beans & peas	uni	uni	uni (2)	uni	8 (21)	13 (53)	15 (76)	7* (55)*	uni (4)	uni	uni	uni
Total area (ha)	*1.26* δ=*1.1*	*1.23* δ=*0.9*	*1.05* δ=*0.6*	*-0.11*	*1.75* δ=*1.4*	*1.76* δ=*1.3*	*1.63* δ=*0.9*	*-0.12*	*1.22* δ=*0.6*	*1.19* δ=*0.7*	*0.89* δ=*0.5*	*-0.33*

Source: Own computation

Notes: (uni) signifies that the crops are unimportant in area coverage (but not equal to zero) and in number of growers. D- Difference in area planted and number of growers between 2001 and 1994.

* indicates differences of paired-samples t-test for mean area cropped and proportion of growers significant at 5% level. Figures in parentheses are the percentage of sample farmers growing the crops. Note that land cultivated can be equal or greater than total land owned as farmers can rent and/or share in land from others.

In conclusion, the increasing percentage of growers and increase in the area under cultivation of tef and maize (and the reduction in the area of the predominant crop, sorghum) in Ankober-Dinqi points a tendency of farmers to diversify their production. A similar explanation can be given for the decline in the area of the dominant crop, barley, and its replacement by wheat and beans/peas in Debrebrehan. On the other hand, the shift of a considerable number of farmers in Shashamane from production of barley and tef to further expand production of the already dominant crops, wheat and maize, suggests tendency on the part of farmers to specialize in these crops. On the whole, the increase in proportion of area and number of growers, particularly for wheat and maize, in all locations is an indication of the increase in the volume of production of these crops relative to the volume of the other crops in the study period. As mentioned earlier, this could be due to advantages from technological improvements in maize and wheat. This is discussed extensively in the next section.

16 Information for the year 1999 was purposely excluded from the Debrebrehan and Shashamane column simply to make the table less bulky and to increase its legibility; the data for that year has no drama and are consistent with those of the other years.

4.5 Analyses of Changes in Technological Progress, Technical Efficiency and Total Factor Productivity

In this chapter, the econometric estimations and economic interpretations of the empirical results of the stochastic frontier models are presented. The chapter is divided into three parts; first, analysis of changes in production technology, and second, changes in efficiency in a specified time. Third, these results are then further used to compute a Malmquist TFP index. Looking at the development of TE, technological progress and TFP by crop and location over the period analyzed, we tried to summarize the trends and the potentials of agricultural transformation in Ethiopia, other factors being constant. Before discussing details of the estimations and the results, the functional form of the stochastic frontier model and a statistical description of the variables entered into the model are presented in brief.

4.5.1 Functional Form of the Stochastic Frontier Model and the Variables

Based on the dynamic theory of optimizing behavior of producers (firms), it is assumed that changes in economic or policy variables will result in a new environment that may, in turn, lead to new optimal decisions and a new economic structure. Farmers' inefficiency may change over time as a result of incentives obtained from the reforms, better access to information, and changes in knowledge of improved input uses due, for example, to better extension services. Differences in the motivation and ability of farmers to use best techniques in the production process can be examined using a random coefficient framework. Therefore, although individual farmers' production behavior is different and they may produce different levels of output using similar levels of technology and input, the fixed coefficient econometric model was found to be less important.

Choosing the functional form of a stochastic production frontier is a serious task of econometric specification. The translogarithmic (transcendental logarithmic) and Cobb-Douglas (CD) stochastic production functions are the two most popular for representing the production process to evaluate agricultural performance. Both functional forms are widely used in modeling agricultural frontier production function (Coelli, 1996). It is argued that these functional forms have peculiar properties that may be useful to consider before choosing for practical applications.

The translog production function has a preferential characteristic in that it is a less restricted (unrestricted) form, allowing more flexibility in the elasticity of substitutions between factors than the CD functional form (Piesse and Thirtle, 2000; Abdulai and Huffman, 2000). However, it is often criticized because it does not yield coefficients of plausible sign and magnitude, mainly due to serious problem of multicollinearity (Fan, 1991; Battese and Broca, 1997). Diewert and Wales (1987) also point out that the translog functional form often fails to satisfy a minimum priority requirement of basic production behavior than the CD function.

The CD function has been widely and extensively used in stochastic frontier production function analysis. It is simple to use but is said to be less flexible (it imposes a constant factor share over time). However, Salim and Kalirajan (1999) argue that it is less restrictive when all the coefficients are allowed to vary. Moreover, the CD function can afford maximum flexibility in dealing with data imperfections (Tybout, 1990). It has been used in many empirical studies, particularly those related to agriculture in developing countries and in transition agriculture. Voigt and Uvarovsky (2001) indicate that the CD function gives more realistic information in describing production behavior in developing countries and transition economies. Thus, the frontier models estimated are defined as:

$$Y_{it} = \beta_o + \sum_{J=1}^{n} \beta_j X_{ijt} + (v_{it} - u_{it}), \quad \textit{Cobb-Douglas function,} \tag{4.9}$$

and

$$Y_{it} = \beta_o + \beta_i \sum_{i=1}^{n} X_{it} + \sum_{i=1}^{n}\sum_{k=1}^{n} \beta_{jk} X_{jit} X_{kit} + \sum_{i=1}^{n} \beta_{jj} \frac{1}{2} (X_{ijt})^2 + (v_{it} - u_{it}) \quad \textit{Translog function} \tag{4.10}$$

where the subscripts i and t represent the i-th farmer and the t-th year of observation, respectively.

In an economy where the government is implementing stepwise structural reforms, sectoral policy decisions such as those relating to agriculture are often dynamic. In addition, producers allocate their resources based on their expectations over several production periods. Therefore, a model aimed at evaluating the performance of firms or sectors needs to be dynamic as well. Accordingly, Cornwell, Schmidt and Sickles (1990) and Battese and Coelli (1992) suggest a time-varying firm effects model for panel data, such that the TE of firms monotonically changes over time. Thus, as the basic idea was to specify a model that is flexible over time, and a linear as well as a quadratic time variable was included in the model to allow shifts in the production function over time owing to technological change. In addition, the stochastic frontier model includes the cross-product of year of observation and the logged inputs to represent the non-neutral technological change. The assumption is that because of neutral and non-neutral technological change, frontier lines shift and change in shape over time, which implies possible changes in marginal products with regard to the inputs used. This gives the model greater flexibility and can be tested for Hicks non-neutral technological change. Technical efficiencies are also measured with regard to the frontier technology level in each given year and can be tested for significant difference over time.

Allowing flexibility of the specification of technical change is very important in that it enables us to separate the effect of technological change from that of TE, which has a different interpretation. TE is a relative measure of management capability given technology. Therefore, an increase in TE implies improvement in the managerial abilities of a firm, which in turn is a result primarily of institutional changes, experience and education, etc. On the other hand, technological change leads to a shift in the production frontier, and derives primarily from investment in research and then adoption of improved varieties and practices. The magnitude and direction of change in TE and technology, therefore, determine the level of total factor productivity. Thus, a flexible model like this is to be recommended for a sector that is relatively unstable and is in the process of transformation (e.g. agriculture in developing countries, where productivity of inputs is expected to change over time owing to expected technological and efficiency changes).

At the outset, the functional form of the stochastic frontier was determined by testing the adequacy of the CD relative to the less restricted translog functions. Thus, a series of hypothesis tests based on generalized likelihood ratio (LR) tests were performed to arrive at a statistically plausible specification of the final model. The LR tests indicated (not reported) that the CD functional form for the data set for all crops was not rejected at a 5% level of significance. The signs and test statistics of the parameters estimated by the translog model were found to be less realistic. This may have been either because the model needed more adequate data, or because it was confronted with many variables, which lead to multicollinearity problem. Therefore, the stochastic CD production function was found to suit our data set better. In this study, the most general functional form of the stochastic frontier for sample farmers in the aforementioned districts was the CD logarithmic function. Accordingly, the functional form of the CD employed in the analysis is defined as follows:

$$\ln Y_{it} = \beta_0 + \sum_{i=1}^{n} \beta_i \ln X_{it} + \sum_{i=1}^{n} \beta_{it} (\ln X_{it} \times t) + \beta_t t + \beta_{tt} t^2 + (v_{it} - u_{it}), \ \mathrm{t = i \ldots n} \qquad (4.11)$$

In a poorly developed market economy, prices do not reflect the actual scarcity and values of resources due to market failure. In such cases, empirical production analysis of physical quantities can explain the performance of traditional and transitional agricultural production more adequately. Consequently, we elected to use the physical quantities of all dependant and explanatory variables in this study. The variables are described as follows:

$\ln Y_{it}$ represents logarithm of quantity of grain harvested (in kilograms);

$\ln X_{it}$ represents logarithm of essential inputs used in the production process in *t*-th year. The inputs considered were:[17]

- X_1 is area cultivated for the crop in hectares,
- X_2 man-day equivalent of family members to proxy total labor used in the production of the crop,
- X_3 total fertilizer in kilograms,
- X_4 number of oxen/bulls owned to proxy animal labor.[18] Oxen are an important source of labor power in traditional agriculture. Moreover, they are considered part of farmer's capital stock.
- X_5 dummy variable (D) was included to capture productivity differences explained by ecological and socio-economic differences as represented by three districts namely Ankober-Dinqi, Debrebrehan and Shashamane. The dummy variable D has the value one if Shashamane district and otherwise zero. Note that Shashamane and Ankober-Dinqi were compared for maize and tef, and Shashamane and Debrebrehan for wheat.

t represents year of observation, where t = 1, 2, 3, and 4 for the year 1994, 1997, 1999 and 2001, respectively.

v_{it} and u_{it} are the random variables as defined above. β_o and β's are parameters of production functions to be estimated.

As has already been discussed, the inclusion of a time variable represented by "year" in the model is intended to capture the inter-temporal technological changes in the farmers' production process. The non-neutral technological change is present if the coefficient of β_{it} is significantly different from zero; neutral technological change exists if β_t and/or β_{tt} are different from zero. However, if all the coefficients are not different from zero, i.e., $\beta_{it} = 0$, $\beta_t = 0$ and $\beta_{tt} = 0$, then there would be no technological change in the production of that particular crop in the specified period and location. In other words, there would be no technological change in the farm production if the coefficients of all variables for the years of observation are zero. Thus, the coefficients β_{it}, β_t and β_{tt} stand for total productivity growth that can be ascribed to technological

---

[17] Data on other intermediate inputs like seed, herbicides and pesticides were not available. Weather conditions (rainfall) and price variables were not included in the model as farmers in a certain community are expected to receive similar rainfall and prices, and hence there would be no variability in short-time panel data.

[18] It is often difficult to obtain quantified data for all inputs used in a subsistence farming system like Ethiopia. For instance, labor time for draft power could not be available. Thus, the number of oxen/bulls owned was used as a proxy to substantiate the validity of estimation of the production function.

change over time, and β_o is the intercept and stands for initial total factor productivity.

The main hypothesis is that there were no changes in technical efficiency and technology over the study period. Following this hypothesis, the estimated models represent *Time-Varying Inefficiency Effects with Technological Change* in the production process of the sample farmers in the study area over the period considered.

4.5.2 Statistical Description of the Variables Entered in the Model

Average trends in quantities of inputs used, output produced and yield of the major crops over the study period are shown in Table 4.11, and the details across the different districts are shown in Tables A4.5, A4.6 and A4.7 in the Annex. Overall, the change in volume of chemical fertilizer used was considerable, but changes in all other inputs were not substantial. Nevertheless, a considerable discrepancy emerged among sample households as regards use of this input. For instance, while sample farmers in Shashamane district most intensively applied fertilizer on all crops, sample farmers in Ankober-Dinqi applied only a little on tef and nearly none on maize. The increase in the area under maize in Shashamane, and under wheat in Debrebrehan and Shashamane (except in 1999) was particularly important. The area allocated for tef seems fairly similar over the period in both Ankober-Dinqi and Shashamane locations. Except in Debrebrehan, draft power per sample household also increased to some extent.

In general, grain outputs per sample household of the three major cereals tef, wheat and maize, increased over the study period. The most spectacular increase in output was recorded in maize production. On average, maize output more than doubled in 2001 compared to the year 1994 (140%) (it increased from 235 to 569 kg), and in the same period yields were 2.7 times higher on average. The annual average growth rate of maize output was about 18%. Given the differences in potential between the districts, the growth in maize output was markedly different as well. The increase in Shashamane was particularly high, around triple, whereas in Ankober-Dinqi it increased only by 1.6 times. Average maize yield was also much higher in Shashamane than in Ankober-Dinqi. The output dispersion among the sample households (indicated by the SD) and the yield gap between the two districts increased. Therefore, there seemed to be increasing disparities in maize production among households and between the districts (see Annex Table A4.7).

In absolute terms, the change in wheat output came second to maize output. Compared to the base period wheat output increased by about 208 kg per household in 2001. This represented an increase of about 62%, and its average annual growth rate hit about 9.6%. Unlike the case of maize, wheat output increased much more in Debrebrehan than in Shashamane. While wheat yield in

Debrebrehan increased by 576 kg (from 275 kg in 1994 to 851 kg in 2001), in Shashamane in the same period it increased only by 25 kg. Likewise, there was little change in the output of wheat in the latter location over the period under consideration, but output and yield were already comparatively high there. Thus, growth in this variable could be exclusively attributed to sample wheat farmers in Debrebrehan. Therefore, contrary to maize production, wheat production between the locations was likely to converge, leaving aside all the differences among the sample households.

Although the magnitude of change in tef output is relatively the lowest, it was approximately double (90%) in 1999 compared to 1994. It attained an average growth rate of about 15% per annum. However, variations in tef output among the sample households increased over time. Such growth in tef output was manifested much more in Ankober-Dinqi than in Shashamane. Similarly, tef yield increased much faster in Ankober-Dinqi than it did in Shashamane. Towards the end of the study period, the increase in tef yield in Ankober-Dinqi was about twice that of the increase in Shashamane. Similar to wheat production, the gap in tef output and yield between the districts was slightly converging.

In summary, an absolute increase in output of all cereals was evident. The extent of the increase in output of the cereals was not, however, uniform across the study areas. The growth of tef output in Ankober-Dinqi was particularly high, while in Shashamane it was maize and in Debrebrehan wheat. Thus, given the initial production potential of the locations considered, the production practice of wheat and tef likely reduced the divergence between the locations under the study. However, the production practice of maize increased the discrepancy. The latter may suggest specialization in crops where the potential for improvement is higher.

It was also observed that in the analyzed period, output of all the major crops considered increased dramatically without a marked change in inputs except chemical fertilizer. So, the causes of such increase in output should be investigated. In the next section, we try to explain the effect of farmers' internal and external governance on production response in the post-reform period in terms of changes in technical efficiency and technological progress using the stochastic production frontier model. The results would also reveal differences within and between the locations, which are marked by distinct differences as regards market accessibility and agricultural potential.

Table 4.11: Summary of Mean Quantities of Inputs and Output in the Stochastic Frontier Model for Tef, Wheat and Maize Farmers

Crop	Inputs	1994	1997	1999	2001
Tef					
	Land (ha)	0.415	0.451	0.423	0.415
	Labor (MD)	3.11	3.65	2.87	3.75
	Fertilizer (kg)	16.65	29.25	45.24	49.80
	Oxen/Bulls (number)	1.06	1.52	1.77	1.85
	Output (kg)	140.74	171.08	267.71	264.00
	Yield (kg/ha)	340.00	379.00	633.00	636.14
Wheat					
	Land (ha)	0.50	0.65	0.48	0.60
	Labor (MD)	3.30	3.42	3.01	3.52
	Fertilizer (kg)	30.10	33.85	44.73	68.43
	Oxen/Bulls (number)	1.97	2.32	2.21	2.42
	Output (kg)	327.98	422.44	472.66	529.50
	Yield (kg/ha)	656.00	650.00	985.00	883.00
Maize					
	Land (ha)	0.322	0.440	0.333	0.361
	Labor (MD)	3.78	3.97	3.16	0.35
	Fertilizer (kg)	34.61	15.31	24.54	41.62
	Oxen/Bulls (number)	1.63	1.25	1.68	1.83
	Output (kg)	234.67	494.80	542.15	569.00
	Yield (kg/ha)	729.00	1125.00	1628.00	2002.77

Source: Own calculation; MD is man-day equivalent

4.5.3 Econometric Estimations and Results

Following the main research objectives, in this section we try to ascertain the extent of changes in production and productivity of major crops after the reform. Assuming no significant change occurs in the natural conditions over the period analyzed the efficacy of development strategies, market liberalization and other institutional changes determine the development of production and productivity of agricultural production through their effects on TE and technological progress. In this section we focus on analyzing the production process using the stochastic frontier model and try to give explanations of the results of estimations in detail. Analysis of changes in technical efficiency, technology and total factor productivity for the major crops selected forms the core of the discussion.

The maximum likelihood (ML) estimates of the parameters of the stochastic frontier production function defined by equation (4.11), given the specification for the technical inefficiency effects defined by equation (4.2), were applied for tef, wheat and maize farmers separately. The parameters of the frontier production function and the inefficiency model were estimated using the software package FRONTIER 4.1. The software first of all provides results of the traditional average response function, the ordinary least square (OLS), and then the ML estimates, together with predictions for efficiency of all firms in the period in which they are observed in the panel survey (Coelli, 1996). The OLS estimation does not consider the effect of the one-sided inefficiency disturbance term u_i and assumes that all the coefficients and the disturbance term are normally distributed. This implies that, if inefficiency prevails in the model, all the β_i estimators (except the intercept) obtained from the OLS are usually consistent but not efficient.

Table 4.12: OLS Estimation of Cobb-Douglas Production Function for the Major Crops

Variables	Parame-ters	Major crops					
		Tef		Wheat		Maize	
Constant	β_o	4.297	5.79***	3.592	10.20***	4.408	5.85***
District	β_1	0.256	1.78*	2.220	10.39***	1.212	5.89***
Larea	β_2	-0.359	-0.45	0.438	3.17***	0.509	1.54*
Llabor	β_3	0.556	1.34	0.422	2.17***	0.353	3.25***
Lfertilizer	β_4	-0.052	-0.64	0.031	0.71	-0.116	-1.16
Loxen	β_5	0.289	2.21**	0.351	2.11**	-0.128	-0.80
year	β_6	-0.318	-0.54	0.851	3.81***	-0.065	-0.10
District×year	β_{16}	-	-	-0.244	-1.62*	0.013	01.02
Larea×year	β_{26}	-0.272	-1.58*	0.433	5.50***	0.054	0.35
Llabor× year	β_{36}	0.065	1.66**	0.119	2.50**	-	-
Lfertilizer×year	β_{46}	-0.090	-1.56*	0.002	0.07	0.086**	1.85
LOxen×year	β_{56}	0.548	1.32	-0.049	-0.72	-	-
(year)^2	β_{66}	0.194	1.58*	-0.042	-1.12	0.093	0.57
(Sigma)^2	σ^2	0.388	-	0.4797	-	0.420	-
Log likelihood function		-7.067	-	-64.352	-	-93.30	-

Source: Own computations from survey data

Notes: ***, ** and * indicate statistical significances of t-values at or better than the 1%, 5% and 10% confidence levels, respectively.

Table 4.13: Maximum Likelihood Estimates of Cobb-Douglas Frontier Production Function for the Major Crops

Variables	Parame-ters	Major crops: Tef		Wheat		Maize	
Stochastic Frontier:							
Constant	β_0	5.742	8.39***	3.991	8.69***	4.862	6.96***
District	β_1	0.316	2.43***	2.1765	9.84***	1.220	6.43***
Larea	β_2	-0.203	-1.01	0.418	2.87**	0.503	1.69**
Llabor	β_3	0.830	2.06**	0.467	2.34*	0.225	3.26***
Lfertlizer	β_4	-0.049	-0.63	0.0304	1.56*	-0.021	-1.36*
Loxen	β_5	0.324	1.06	0.326	1.97**	0.053	1.83**
year	β_6	-1.456	-2.81***	0.6976	2.89**	-0.716	-1.05
District×year	β_{16}	-	-	-0.425	-5.31***	0.032	0.1.42*
Larea×year	β_{26}	0.224	2.07**	0.127	2.51**	0.058	1.41*
Llabor×year	β_{36}	-0.271	-2.06**	-0.133	-1.87**	-	-
Lfertlizer×year	β_{46}	0.065	1.93**	0.0038	0.225	0.082	2.10**
LOxen ×year	β_{56}	-0.073	-0.55	-0.0374	-0.66	-	-
(year)^2	β_{66}	0.507	4.58***	-0.0263	-0.68	0.271	1.78*
Variance parameters:							
(Sigma)^2	σ^2	0.246	7.12***	0.460	8.40***	0.676	2.14*
Gamma	γ	0.041	1.47*	0.0741	1.63*	0.510	2.96***
Mu	μ	-		-0.3759	-0.74	-	-
Eta	η	3.405	5.51***	0.504	2.23**	-0.024	-1.35*
Loglikelihood function		-119.7	-	-44.23	-	-82.75	-

Source: Own computation from survey data

Notes: ***, **, and *: Significant t-values at or better than the 1%, 5% and 10% confidence levels, respectively.

On the other hand, the ML estimate takes into account the existence of inefficiency disturbance and thus computes parameters σ^2, γ, μ, and η that help us understand the extent of technical efficiency and trends in the level of technical efficiency. The relative importance of the OLS and ML estimations, therefore, depends on the presence of significant technical inefficiency in the production. Thus, it is essential first of all to undertake a statistical test that would enable us to select a model that is more appropriate for the data. Table 4.12 and Table 4.13 respectively report coefficients of the OLS and the ML estimates of the stochastic CD production functions.

4.5.3.1 Hypotheses Test

Given the specification of the stochastic frontier production model with time-varying inefficiency effects, and based on the results of the models, the outcome of different hypotheses tests are presented in Table 4.14.

Table 4.14: Summary of Tests of Hypotheses for Parameters of the Stochastic Frontier Production Function

No.	Null hypothesis	Description	Decision by crop farmers		
			Tef	**Wheat**	**Maize**
1	OLS adequately represents the data	LR (λ) $=0$	Rejected	Rejected	Rejected
2	Farmers are fully technically efficient	Ho: $\gamma=0$	Rejected	Rejected	Rejected
3	Technical inefficiency effects are time-invariant	Ho: $\eta=0$	Rejected (improved)	Rejected (improved)	Rejected (deteriorated)
4	No technological change	Ho: $\beta_5=\beta_{i5}=0$	Rejected (regressed)	Rejected (progressed)	Rejected (progressed)
5	No change in productivity difference between regions over time	Ho: $\beta_{16}=0$	Accepted	Rejected (decreased)	Rejected (increased)

Source: Own compilation from analysis of the frontier function

Hypothesis 1: The first hypothesis was that the traditional mean response function is an adequate representation of the data (that is, Ho: $\lambda = 0$). The LR[19] tests were conducted for each model (represented by crop) to determine whether the OLS can represent the data as adequately as the stochastic frontier does. The LR computed value for tef was 14.72, for wheat 40.24 and for maize 21.10, with two restricted variables for tef and maize, and three restricted variables for wheat. The critical LR values at the degrees of freedom given by the number of variables restricted are: 12.51 for tef and maize, and 7.12 for wheat. The results of the LR tests therefore signify that the OLS representation was strongly

[19] In LR statistics, λ is defined by $\lambda= -2\ln (LHo)/L (H_1) = -2(\ln LHo-\ln H_1)$, where Ho and H_1 are the null and alternative hypotheses involved, with the degree of freedom of parameters restricted. If the null hypothesis, Ho, is true, then λ is asymptotically distributed as a chi-square (or mixed chi-square) random variable. If the null hypothesis involves $\gamma=0$, then λ has a mixed chi-square distribution (see Coelli, 1995; 1996).

rejected for all crops, implying that the ML estimates of the stochastic frontier model fitted the data better. It is also evident that, in the results of the ML estimates, more parameters of the explanatory variables have become statistically significant than in the OLS estimates. Thus, the ML estimates of the stochastic frontier model are preferred.

Hypothesis 2: The second hypothesis was that the variance in the technical efficiency effects are simply random errors and so those farmers are technically fully efficient (that is, Ho: $\gamma = 0$). As noted earlier, the magnitude of the parameter of γ indicates the presence of technical inefficiency. If γ is closer to zero, technical inefficiency is not important; that is, farmers are nearly fully efficient, in which case the OLS can also adequately represent the data. But the closer γ is to unity, the more likely it is that the frontier model is appropriate. In our analysis, the γ coefficients were found to be statistically significantly greater than zero at least at the 5% level for all the sample farmers growing the three major crops. This indicates that the variation in the actual and maximum (frontier) output of farmers producing the crops considered also arose from differences in farmers' practice. The coefficients of γ for tef and wheat farmers indicate that error due to inefficiency on average amounted to only about 4% and 7.4%, respectively. That is, a large proportion of the total variability in output of these two crops was explained not by technical inefficiency, but by white random shock effects, which were beyond the control of the farmers. Nonetheless, over 50% of the total variability in maize output was due to technical inefficiency effects of maize farmers. This can be considered a large loss.

The distribution of inefficiency effects in the production of the crops was not uniform. For tef and maize farmers it was half-normal, but for wheat, distribution was half-truncated. In general, this suggests that tef, wheat and maize farmers in the study area were not fully technically efficient over the study period. This result corroborates the LR test that OLS is not an adequate means of representing the data.

Hypothesis 3: The third hypothesis was that farmers' inefficiency effects are time invariant (i.e., Ho: $\eta = 0$). As discussed earlier in the methodology section, the parameter η measures changes in inefficiency over time. The result shows that technical inefficiency effects significantly tended to decline systematically for tef ($\eta = 3.45$) and for wheat ($\eta = 0.504$), since the parameters were positive and significant (at 1% level) for these two crops. This suggests that their TE significantly improved over the study period. For maize production, the data also rejected the null hypothesis at 10% level. However, the negative sign of the parameter ($\eta = -0.024$) indicates that the technical efficiency of maize farmers declined in the analyzed period. This issue was of such importance that we could not afford to ignore it. Thus, the Technical Efficiency Effect model developed by Collie (1995) was applied to identify the determinants of technical inefficiency in maize-producing farmers (see Chapter 5).

Hypothesis 4: The fourth important hypothesis was that no neutral and/or no non-neutral technological change prevails in production of the crops (that is, Ho: $\beta_6 = \beta_{i6} = 0$, i=1, 2, ..., 6). For all the three crops considered, analysis of the data did not support the null hypothesis. Coefficients of the time variables (neutral change) and of time-input interaction of variables (non-neutral change) appeared to be significantly different from zero at least at 10% level. This implies that the isoquants/frontiers shifted significantly and changed their shape over the period 1994-2001. Consequently, it was evident that technological change occurred in all production lines. However, the magnitude and the direction of the changes were different, determined by the sign and size of the coefficients of the estimates. There appeared to be technological progress in wheat and maize production but technological regress in tef production.

Hypothesis 5: The last hypothesis refers to whether the productivity gap between the regions under study is considerable and significantly changes over the study period. The district dummy variable included in the model represents the relative productivity difference between the sample locations, mainly due to differences in natural endowments such as soil quality, climatic suitability etc., and socio-economic variables. The parameter (β_1) was significant at the 1% level in production lines. The positive and significant value of the district dummy for tef, wheat and maize indicates that the productivity of these crops for farmers in Shashamane was relatively higher than for sample farmers in Debrebrehan and Ankober-Dinqi. As described earlier, a similar result was documented in the descriptive analysis of the crop yields across the districts. Thus, the econometric result confirmed the superiority of the crop yields in Shashamane compared to the other locations.

The time-cross effect of district dummy (β_{16}) was not meaningful for tef, and hence was excluded from the model. This implies that the productivity differential between the locations studied did not significantly change in tef production over the study period. However, it was positive (significant at 10%) for maize farmers but negative (significant at 1% level) for wheat farmers. This implies that while the wheat productivity gap between Shashamane and Debrebrehan decreased in the study period, the productivity gap in maize production between Shashamane and Ankober-Dinqi increased. The implication is that differences in resource productivity were increasing in maize but diminishing in wheat production in the regions considered over the period analyzed. The convergence of productivity of wheat in these two districts could be due to the increasing use of improved varieties of wheat and fertilizer in Debrebrehan, given its initial potential. The farmers in Shashamane already had more experience in the use of such inputs.

4.5.3.2 Changes in Output Elasticities and Technological Progress

Elasticities

As the ML estimates of the frontier CD production better represented the data for all sample growers of the selected crops, the remaining part of our discussion is based on the results of this model presented in Table 4.13. The estimates of the frontier coefficients signify the maximum possible contribution of the inputs to output, when firms are operating on their frontier by adopting best practice techniques. The estimates of the variables and their interactions with time have sensible values and acceptable signs. The signs and the magnitudes of the coefficients of the variables conform to theoretical expectations. Note that in the maize production model, only the cross-time effect of land and chemical fertilizer were included; the time interaction of labor and oxen was excluded, as these were not found to give sensible information.

Partial output elasticities with respect to the input variables are the most important tool to describe production technology. The output elasticity explains the extent to which a proportional change in an input is required to change output by a certain proportion, where the other factors are held constant. The output elasticities can be estimated at the mean output with respect to the k-th input variables in year t for each study period (Lundvall and Battese, 1998). Therefore, from the specification of the CD stochastic frontier production function (4.11), the output elasticities with respect to each input can be mathematically derived as:

$$E_{Xk} = \frac{\partial \ln \left[E\left(Y_{it}\right)\right]}{\partial \ln X_{k}} = \beta_{k} + \sum_{j}^{n} \beta_{it} t \tag{4.12}$$

Given the specification of logarithmic CD production function and the empirical results indicated in Table 4.13, the partial elasticities of frontier output (best practice) with respect to each of the inputs land, labor, fertilizer and oxen power are presented in Table 4.15.

One of the basic regularity conditions of the production function is *monotonicity*. This implies that an additional unit of any input should not decrease the level of output, that is, marginal productivities are positive. Such a theory explicitly excludes the third, irrational stage of the production function, where the marginal productivities are negative (Coelli, Rao and Battese, 1998). Thus, the production function is monotone if the production elasticities (E_i) with respect to inputs (X_i) are non-negative. That is,

$E_i \geq 0$, for input i = X_1, X_2, X_3, and X_4.

In the earlier periods, in 1994 and 1997, the traditional inputs land, labor and oxen had the largest contributions (elasticities) as a whole. Land accounted for most of the output growth in all production lines throughout the study period. In 1994, land accounted for over 50% of the variance in the output of wheat and maize but only about 2% for tef. In the same year, labor explained about 56% in tef, 33% in wheat and 22% in maize. Oxen accounted for over 25% of the total variation in tef and wheat outputs. However, the relative importance of labor (except in maize) and oxen generally decreased. Eventually, labor played a decreasing role in production of tef and wheat. Meanwhile, the contributions of chemical fertilizer[20] and land[21] were rapidly increasing for all crops. For example, in 2001 land accounted for over 65% of the total variation in output of all crops, and fertilizer for about 30% in maize, 20% in tef but only 5% in wheat output. However, in the study districts, there was little chance to increase land size cultivated in the study period. The only way to increase the area for one crop is by reducing that of other crops, or that of grazing land, if any.

Although the contribution of labor remained important for maize production, its relative importance in the production of tef and wheat was declining and eventually became negative,[22] violating the assumption of monotonicity. The negative value of production elasticity with regard to labor for tef and for wheat implies that employment of additional labor is irrational. Similar results are documented in Curtiss (2002) for Czech crop production.

This can be interpreted using the concept of "surplus labor", where the marginal product of labor becomes zero after a certain level of output; it relates even more closely to the concept of the "hard-working peasant" (Mellor, 1985), where the marginal product of labor becomes negative. Mellor argues that the latter might be typical in large parts of Sub-Saharan Africa: because of limited arable land, but increasing population pressure in the rural sector, the average and marginal product could be just equal to the subsistence wage (the amount

---

[20] This is with the exception of fertilizer for wheat that arose perhaps due to intensive and probably similar rate of application.

[21] From the CSA series data, annual growth in output at macro level is largely associated with an increase in the area of land cultivated. However, this must be solely due to encroachment on forest and bush or marginal land by the inhabitants of the peripheral part of the country. In the central part of the country (such as the area where the study was carried out), there is little scope for doing this, and the available land has been continuously sub-divided because of ever-increasing population pressure. Therefore, the increasing elasticity of land over time should be due to an increase in the marginal productivity of land resulting from increasing use of land-augmenting technologies, particularly improved seed and fertilizer.

[22] Based on the results in Table 4.2, all inputs except labor have the desired sign and are significantly different from zero. There appeared very little hired labor, but family labor inputs may be excessive. The negative elasticity of frontier output for labor indicates that labor use in the study areas was not optimal.

required for one person to exist); no surplus output will be forthcoming. The implication is that once the maximum output level is reached, the farm cannot support any further increase in labor. Consequently, additional labor is portrayed as operating in the third stage of the production function in which the marginal product of labor becomes negative. Mellor (1985) underlined that this is actually an indication of the incidence of poverty. This is the case in which Lewis (1954) and Fei and Ranis (1964) argue for the transfer of labor from agriculture to industry using the dual economy model, although the theoretical and empirical applications of this model remain controversial. In the short run, it may be advisable to reallocate labor to other crops (like maize) for which its marginal productivity has been positive. In the long run, creation of non-farm employment opportunities and generation of labor-intensive technology (for tef and wheat) may be the most feasible solutions.

The other interesting question may be: what explains the differences in the structure of partial output elasticities for the different lines of production? The most plausible reason could be differences in production technology. The low output elasticities with regard to labor for tef and wheat imply that production of these crops was relatively more labor-intensive, tef being the most labor-intensive. But labor input played a more significant role in maize than in tef and wheat production, and the output elasticity in respect of land was the highest. This could be mainly due to the availability of land-augmenting technology (such as improved seed) for wheat and maize. The proportional increase of land resulted in a higher proportional increase in maize and wheat than in tef.

Technological Progress

The change in production elasticity over time should, by definition,[23] be the joint effect of (i) the change in marginal productivities due to change in the quantity of inputs used, and thus movements along the production curve by increasing level of inputs (scale of production); and (ii) the change in the inverse average productivities of the utilized inputs due to the upward or downward shift in the production curve resulting from technical change. As noted before, the time and cross-time interaction of certain input variables included in the models (equation 4.11) is intended to capture the relative effect of technological change on productivity. The input-time interaction variables tell us that the partial output elasticities with respect to those inputs do vary over time owing to Hicksian

[23] The general formula of output partial elasticity for output Y and input X can be calculated as: $E_x = \frac{\partial Y}{\partial X} \times \frac{X}{Y} = MPP_x \times \frac{1}{APP_x}$

non-neutral technological change. Production elasticity of input i in these production functional forms increases if $\beta_{it} > 0$, but decreases if $\beta_{it} < 0$.

The average measure of technological change can be obtained from the first derivative of the estimated function with respect to time t at mean values of the inputs used in each year. This can be computed as:

$$\frac{\partial \ln[E(Y_{it})]}{\partial t_t} = \beta_t + 2\beta t_i + \sum_{j}^{n} \beta_{it} X_i \quad (4.13)$$

In the estimated models (Table 4.13), the neutral technological change is explained by the aggregated effect of coefficients of β_6 and β_{66}. This could be represented by the sum of the first two terms in equation (4.13). If the sum is significantly greater than zero (positive), then there is neutral technological progress, and the reverse implies technological regress. On the other hand, the non-neutral Hicksian technological change is explained by coefficients of time-input interaction represented by coefficients $\beta_{26}\ldots\beta_{56}$ in the same Table. The last term in equation (4.13) represents the magnitude of changes. Essentially, the total change in technological progress in the process of production is determined by the cumulative effect of both the neutral and non-neutral technological changes.

Although the time considered is short, technological change, given the data in the three production lines, was significant. Technological change in wheat and maize was positive, implying technological progress. However, despite the theoretical expectation of positive technological change, the result shows technological regress in tef production over the period analyzed. This may be surprising. In particular, the large negative value of β_6 and the relatively smaller positive value of β_{66} suggest neutral technical regress in tef production. In other words, the production frontier of tef shifted downward, possibly as a result of limited scope of technological choice and land degradation.

The neutral technological change in wheat and maize could be the result of adopting improved varieties and improved agronomic practices, soil management, etc. The non-neutral technological change could be the result of more proportional use of chemical fertilizer and land inputs over time. For example, the significant positive coefficients of land-year (β_{26}) and fertilizer-year (β_{46}) indicate technological progress in all crops considered through increasing use of land and fertilizer. On the other hand, the negative coefficient of the labor-year and oxen-year variables (β_{36} and β_{56}) implies that increasing use of labor and oxen results in technological regress in tef and wheat. In tef production, the negative coefficients of time and the cross-products of time with labor and oxen surpassed the positive effects of the time squared and the area-time interaction.

The Hicksian non-neutral technological change could increase the income gap among the farmers and between regions. If the other factors are held constant, individuals and/or regions that have relatively abundant land and those who can afford to increase their use of fertilizer could possibly benefit much from technological progress in wheat and maize production.

Table 4.15: Summaries of Production Elasticities of Inputs over the Study Period

Crop	Inputs	Year			
		1994	1997	1999	2001
Tef	land	0.021	0.245	0.468	0.693
	Labor	0.559	0.288	0.017	-0.254
	Fertilizer	0.016	0.081	0.146	0.211
	Oxen	0.251	0.178	0.105	0.032
	RTS	0.847	0.792	0.737	0.682
Wheat	land	0.545	0.672	0.799	0.926
	Labor	0.334	0.201	0.068	-0.065
	Fertilizer	0.034	0.038	0.042	0.046
	Oxen	0.289	0.251	0.214	0.176
	RTS	1.202	1.162	1.123	1.083
Maize	Land	0.561	0.619	0.677	0.735
	Labor*	0.225	0.225	0.225	0.225
	Fertilizer	0.061	0.143	0.225	0.307
	Oxen*	0.053	0.053	0.053	0.053
	RTS	0.930	1.070	1.210	1.350

Source: Own computation

Notes: * indicates constant or has no time-interaction effect; technical or technological change is the elasticity of output with respect to time, year.

4.5.3.3 Changes in Returns to Scale

The concept of returns to scale (RTS) is widely used in production and cost functions analysis. It measures the change in output resulting from a proportionate change in inputs (average cost), which in turn depends on the technical characteristics of the production function. The RTS can be categorized as increasing (economies of scale), decreasing (diseconomies of scale) or constant. If a proportionate change in all inputs results in a more than proportionate change in output, then the results is an *increasing* RTS, and the reverse is a *decreasing* RTS. *Constant* RTS occurs when a proportionate change in all inputs produces the same proportionate change in output.

Closely associated with RTS is the concept of *scale efficiency*, introduced by Färe, Grosskopf and Lovell (1985) based on Farrell's (1957) technical efficiency. According to the neo-classical theory of profit maximization under perfectly competitive market conditions, scale efficiency occurs when long-run average cost is minimized (Porter and Schully, 1987) or when an input-output combination results in zero-profit long-run of competitive equilibrium. In relation to production technology, a farm is scale efficient if it displays a constant RTS (Färe, Grosskopf and Lovell, 1985). In the economics literature, a constant RTS is most frequently assumed. However, the optimal size of an operation is related to issues of ownership and property rights, and dynamic development in techniques of production, and thus also motivation and effort issues (Deininger, 1995). Optimal size varies with technological change, since in most cases cost reduction follows technical change, thus resulting in a partial increase in economies of scale (Groenewald 1991).

The magnitude of the sum of the elasticities (ΣE_i) of the mean frontier production with respect to the mean of all the inputs used indicates the type of RTS and hence the extent of scale efficiency. It can be considered as increasing RTS if $\Sigma E_i > 1$, decreasing RTS if $\Sigma E_i < 1$, while RTS is constant if $\Sigma E_i = 1$ provided that it satisfies statistical tests. The statistical tests can be undertaken using the generalized loglikelihood ratio test of linear homogeneity restrictions, $\Sigma E_i = 1$, for input i (land, labor, fertilizer and oxen) for all crops. This can be done by normalizing the output and inputs by dividing them all by one of the inputs. In this study, land was chosen as a denominator to normalize since it is most often chosen for this purpose by many scholars (e.g., Curtiss, 2002). The RTS is constant if the joint effect of all parameters associated with land in the CD functions is equal to zero. Chi-square (χ^2) test statistics were used in the analysis.

Assuming that farmers had enough time and opportunity to vary their factors, the response of output to the scale change was computed. The bottom rows in each category in Table 4.15 show the RTS. The result shows that the assumption of linear homogeneity that states that the production RTS is constant was rejected by the data. Hence, no production line was found to be scale efficient in the study period, except wheat in the last year of observation. The RTS for maize was greater than one (at 10% level) over the study period, indicating evidence of an in increase in RTS. An RTS of more than one implies a proportional increase in all inputs in maize production, resulting in a higher proportional increase in output, and then extension of production would be required to achieve scale optimum. In other words, average maize farmers can still increase use of level inputs to achieve a higher average productivity and production. This movement of maize production away from scale optimum, despite the technological incentives, may be due in particular to lack of financial capital (credit) for the purchase of fertilizer and improved seed, shortage of land for expansion, and lack of market incentives. In this regard, studies in sub-Saharan countries show that farmers apply chemical fertilizer and improved seeds below the recommended rate. Farmers have been constrained by lack of financial resources, absence of adequate supply and instability of product prices (discussed at length in Chapter 5).

Meanwhile, the RTS for tef were generally significantly less than one over the same period. This implies that tef production exhibited a significantly decreasing RTS. This could be due to extra allocation of resources to tef production, particularly labor, as its negative elasticity coefficient indicated. Thus, the result suggests a need to reduce the quantities of input use in tef production. However, farmers might be reluctant to do this because of market advantage (tef is most preferred cereal crop for food particularly in urban hence its price is relatively most stable), and because tef performs better under adverse conditions (it is a risk-minimizing crop).

4.5.3.4 Changes in Technical Efficiency

Changes in TE depend on farmers' internal and external factors. It is most affected by incentives resulting from economic and institutional reforms such as market liberalization, research and extension services and improved information. Institutional reform can induce farmers to link their efforts and resources to better production techniques. In other words, farmers could allocate their time and resources in a more efficient and profitable way. The size of the TE index thus provides information on the potential for improvement in terms of management practices. Having the idea of the crop production technologies described by the frontier output, differences in farmers' ability to achieve the frontier given the available inputs was analyzed on the basis of TE for each crop. Following equation (4.11) above, the maximum possible outputs achievable with the given technology were obtained to estimate the firm-specific TE indices. The technical efficiencies are predicted using the conditional expectation of *exp* $(-U_{it,})$ from equation (4.3 and 4.4), given the composed error term of the stochastic frontier.

Firm-level technical efficiency analysis over the specified period showed mixed results (Table 4.16). While the efficiency of wheat and tef farmers significantly improved, maize farmers – unsurprisingly– became more inefficient. Such changes in TE are actually consistent with the hypothesis tests discussed earlier (Table 4.13 and Table 4.14). Improvement in the TE of tef and wheat-producing farmers was dramatic. With continuous improvement over the period under study, the mean TEs of the farmers growing the two crops rose to a mean of about 0.94 in 2001, which was an increase of more than 20% compared to the base year. Consistent with the theoretical concept of efficiency growth, the improvement in TE for wheat and tef increased at a decreasing rate, and analysis of paired-mean difference revealed the change over the years considered to be significant at 10% and 5% respectively. The rate of progress in TE was noticeably much higher in the initial period compared to the later periods. This implies that farmers responded quickly to the institutional changes and eventually internalized the effects of reform well. Thus, the systematic time-varying inefficiency change could indicate that development in efficiency positively contributed to technical progress in the production of the two crops.

The variability of TE of the tef and wheat farmers (implied by the standard deviation and the ranges) was also greatly reduced in the period analyzed. For instance, in 1994 the range between the most and the least efficient farmers was as high as 64% and 51% for tef and wheat respectively. These gaps had diminished to about 20% in 2001. Compared to 1994, farmers performing at the lowest level of TE also improved: tef farmers from 0.22 to 0.78 and wheat farmers from 0.37 to 0.79 in the year 2001.

Interestingly, as mentioned earlier, the opposite was the case with the maize farmers. In the earlier period, maize producers were relatively even more efficient than producers of the other two crops. Nevertheless, they gradually became more and more inefficient over the period analyzed, and eventually it was found that on average maize was the most inefficient production line of all the crops analyzed. The mean TE of the maize farmers declined from 0.89 in 1994 to 0.76 in 2001, which represents a fall of around 15%. This suggests that in the last year of the study, on average about 24% of the potential (maximum) output was lost owing to inefficiency. In the same period, the lowest efficiency level fell further, from 0.81 in 1994 to 0.55 in 2001 (a 26% fall).

Table 4.16: Summary of Descriptive Statistics of Farmer-Specific Technical Efficiency by Crops over Time

Crop	Description	1994	1997	1999	2001
Tef					
	Minimum	0.22	0.57	0.82	0.78
	Maximum	0.86	0.95	0.98	0.99
	Mean	0.76	0.85	0.94	0.95
	Std deviation	0.15	0.08	0.04	0.03
Wheat					
	Minimum	0.37	0.54	0.68	0.79
	Maximum	0.88	0.93	0.95	0.97
	Mean	0.77	0.85	0.90	0.94
	Std deviation	0.08	0.06	0.04	0.03
Maize					
	Minimum	0.81	0.69	0.55	0.55
	Maximum	0.95	0.92	0.87	0.87
	Mean	0.89	0.82	0.79	0.76
	Std deviation	0.11	0.16	0.18	0.22

Source: Own computation

Notes: TE ranges between zero and one, with one being the best or perfectly technically efficient.

Moreover, the variability of TE of maize farmers increased, confirming the low level of TE. This could be seen in the discrepancies between the maximum and the minimum, or from the standard deviations. This indicates that differences in farmers' management ability in maize production were increasing among the farmers. It may be because maize production has become more complex due to the introduction of new technologies. In other words, higher technological complexity coheres with greater scope for choice, and this provides comparatively more space for farm management failure in production decision-making. This could mean that the decline in TE in maize production, assuming market incentives were unchanged, might be due to the existence of increasingly greater technological differences. As observed during the survey, only a few farmers used improved varieties, and those who did had difficulties and displayed a high degree of variability as regards following the recommended agronomic practices. Only a few of the best farmers were able to cope with rapid technological change, leaving a significant number of farmers behind, which resulted in a decrease in the scale of TE and an increase in its variability (see Chapter 5 for details).

Regional Differences in TE Changes

Table A4.8 in the Annex provides detailed descriptive statistics of TE of farmers for each crop and location over the study period. Although no significant difference in the TE of wheat farmers was observed across the districts, tef producers in Ankober-Dinqi and Shashamane were quite different over the study period. On average, sample tef producers in Shashamane were technically 2.3% to 6.0% more efficient than tef farmers in Ankober-Dinqi (significant at 5% level). However, the difference and the variability between the locations diminished gradually over the study period. Although maize growers in Shashamane were relatively more efficient in the earlier periods, contrary to the case of tef, the TE in this district deteriorated faster than those in Ankober-Dinqi. This may be due to the fact that a greater degree of technological change (in terms of use of improved seeds and improved cultivation practices) was observed in Shashamane than in Ankober-Dinqi, and this could lead to greater management complexity, which in turn could reduce TE. Consequently, in considering the whole period under investigation no real difference was observed in the level of TE of the maize farmers between these two districts.

In conclusion, efficiency measures indicate that the household production system contributed significantly to output growth of wheat and tef, but not of maize. The empirical results for TE indicate that there is considerable potential for increasing the production level of maize from the available resources and technology that was lost because of inefficiency. The efficiency gap in performance among the regions was reduced. However, farmers in Shashamane still showed overall higher TE in almost all production lines. This may be due mainly to the region's natural agricultural potential and the substantial impact of the metropolis on farming here, in other words, the fact that the region enjoys better market access and a better transport infrastructure.

4.5.3.5 Changes in Total Factor Productivity and its Relative Contribution to Output Growth

The decomposition of output growth into its major components provides valuable insights concerning productivity. As mentioned in Section 4.5.2 (description), the absolute output growth in each crop was considerably high over the study period.[24] Such changes in output could be due to changes in all or some of the components, such as increase in input used and/or improvement in productivity, which in turn resulted from improving technical efficiency and technological progress. In the preceding sections of this chapter, we have tried to explain the changes in production technologies based on partial output elasticities and parameters representing the neutral/non-neutral shift of the production frontier. In addition, changes in technical and scale efficiencies have been described. In this section, we have tried to synchronize the aggregate effect of changes in technological progress and technical efficiency on total (overall) factor productivity (TFP).

To evaluate productivity development over the study period, the Malmquist TFP index was used. The Malmquist TFP comprises a technical efficiency change element and technological change element. Coelli, Rao and Battese (1998) show that the technical efficiency of firm *i*, TE_{it}, is a distance function at time t. The efficiency change element of the Malmquist TFP index ($TEC_{i(t+1)}$) is the ratio of two TE distance functions for time t+1 and t based on equation (4.3). This can be calculated as:

$$TEC_{i(t+1)} = \frac{TE_{i(t+1)}}{TE_{it}} \tag{4.14}$$

Similarly, the technological change index ($TCH_{i(t+1)}$) between two adjacent periods, t+1 and t, for firm i, can be directly calculated from the estimated parameters of the stochastic production frontier. This can be done by evaluating the partial derivatives of the production frontier with respect to the time variable, t, for consecutive years, based on equation (4.13). According to Coelli,

---

[24] The notable improvement in output growth in the study period could also have been determined by the relatively good weather conditions (rainfall) and the resultant good harvest. It should be acknowledged as well that the origin of the growth in output of the crops should be complemented by favorable output prices in the specified period.

Rao and Battese (1998), the geometric mean of these calculations gives us the index of technological change for two adjacent periods:

$$TCH_{i(t+1)} = \sqrt{\left\{1+\frac{\partial f(x_{i(t+1)},(t+1),\beta)}{\partial(t+1)}\right\}\left\{1+\frac{\partial f(x_{it},t),\beta}{\partial t}\right\}} \tag{4.15}$$

Following Coelli, Rao and Battese (1998), the product of the indices of TEC and TCH obtained from equations (4.14) and (4.15) is a Malmquist TFP index. This can be described as:

$$TFP_{it} = TEC_{it} \times TCH_{it}. \tag{4.16}$$

The growth of output can then be decomposed into productivity growth, arising from technological and TE changes and changes in the scope of inputs used based on the illustration given in Figure 4.2. Recalling the functional form of the model used (equation 4.11) in the empirical estimation of the frontier analysis, a simplified form of the specification can be rewritten and the contribution of each component to the output growth can be mathematically derived. The computation is derived as follows:

Recalling the functional form of the specification of the stochastic frontier CD production function,

$$\ln Y_{it} = \beta_0 + \sum_{i=1}^{n}\beta_i \ln X_{it} + \sum_{i=1}^{n}\beta_{it}(\ln X_{it} \times t) + \beta_t t + \beta_{tt} t^2 + (v_{it} - u_{it}),\ t = i \ldots n;$$

we can rewrite it as:

Let $\ln\alpha_o(t) = \beta o + \beta_{tt} + \beta_{tt}{}^2 + v(t)$; $\gamma_i(t) = \beta_i + \beta_{it}$; $E(t) = u(t)$

$\ln Y_t = \ln\alpha_o(t) + \Sigma\,\gamma_i(t)\ln X_i(t) + E(t)$; and the growth rate is given as

$\partial \ln Y_t/\partial t$	$= \partial\ln\alpha_o\,(t)/\partial t$	(neutral technological change)
	$+ \Sigma\,\gamma_i\,(t) \times \partial\ln X_i(t)/\partial t$	(input change)
	$+\Sigma\,\ln X_i(t) \times \partial\,\gamma_i(t)/\partial t$	(non-neutral technological change)
	$+ \partial E\,(t)/\partial t$	(technical efficiency change)

Where Y_t represents output and X_t input at time t; and parameters are estimated by the stochastic frontier function. As already mentioned, a separate analysis of the effect of the neutral and non-neutral technological change on productivity did not give a meaningful economic interpretation. Thus, the lump sum effect of both types of technological change, considered as total technological change, was computed in the accounting index. The calculated technological and efficiency change indices, together with the Malmquist TFP indices for the three crops over the period analyzed is presented in Table 4.17. The growth rate of these indices and their relative contribution to the output growth were computed based on equation (4.16) simulating with a linear logarithmic approach for the year 1994 to 2001. Details of differences across the regions (districts) for each crop considered are presented in Table A4.9 in the Annex.

The result shows that growth trends in the TEC and TCH indices for the different crops varied, and the combined impact of changes in TEC and TCH on TFP for each crop was wide-ranging. For wheat production, both TEC and TCH were growing, but for tef and maize production it was different. While TEC for tef was growing, its TCH was deteriorating, while the reverse was the case for maize. As noted earlier, TE for maize growers fell over the study period, and its TEC indices likewise showed a downward trend. However, TCH for maize improved appreciably in line with positive technological change. Consequently, compared to the base year (1994), TEC increased by about 25% for tef and 22% for wheat, but fell by 15% for maize in 2001. And technological change accounted for an improvement of 31% in wheat and 69% in maize, but a deterioration of 36% in tef. Changes in TEC and TCH eventually resulted in TFP growth of 60% in the case of wheat and 40% in maize, but a decline of 20% in tef.

Obviously, the positive changes in both TEC and TCH for wheat production resulted in a considerable growth in TFP. The average annual growth rate of TFP for wheat was estimated to be 8.54%, which resulted from a growth of 3.88% in TEC and 4.66% in TCH. Consequently, growth in TFP contributed about 89% of total annual growth rate of wheat output (9.63%), with the remaining 11% coming from increased use of total inputs. More specifically, about 48% and 40% of the growth in wheat output was attributed to technological and efficiency improvement respectively. On the whole, although the growth rate in at wheat output was not so extraordinary, the contribution of TFP to growth in the output was remarkable. This was not the case with the maize or tef production process.

On the one hand, the fall in TEC in maize was not so great relative to the huge growth in its TCH index to have much negative impact on its TFP and output growth. TEC fell on average at a rate of 2.86%, whereas TCH grew at 10.77% per annum, and the net effect was positive, resulting in a TFP growth rate of 7.91% over the study period. Thus, about 44% of the total output growth for maize (18%), was explained by growth in TFP of maize and the remaining larger share (56%) by an increase in total inputs used (mainly chemical fertilizer).

Table 4.17: Changes in Indices of Technical Efficiency, Technological Change and Total Factor Productivity and their Relative Contribution to Output Growth

Crop type	Source of output growth	Year				Growth rate	Contribution to output growth (%)
		1994 (=100)	1997	1999	2001		
Tef	Efficiency	1.00	1.116	1.236	1.250	3.82	25.71
	Technology	1.00	0.504	0.723	0.641	-5.27	-35.46
	TFP +	1.00	0.561	0.894	0.801	-1.45	-9.75
	Inputs $	-	-	-	-	16.31	109.75
	Output	1.00	1.220	1.902	1.876	14.86	100.00
Wheat	Efficiency	1.00	1.104	1.170	1.221	3.88	40.30
	Technology	1.00	1.371	1.335	1.310	4.66	48.40
	TFP	1.00	1.514	1.562	1.600	8.54	88.70
	Inputs	-	-	-	-	1.09	11.30
	Output	1.00	1.290	1.440	1.620	9.63	100.00
Maize	Efficiency	1.00	0.932	0.898	0.853	-2.86	-15.90
	Technology	1.00	1.273	1.489	1.688	10.77	59.84
	TFP	1.00	1.186	1.337	1.401	7.91	43.94
	Inputs	-	-	-	-	10.54	56.06
	Output	1.00	2.110	2.310	2.420	18.00	100.00

Source: Author's computations
Notes:

+ Indexes of the TFP, for the years specified, are computed based on equation 4.16, and its growth rate is the sum of the growth rate in the technical efficiency and technological progress.
$ Because of lack of full information, it was difficult directly to compute index for inputs used by the sample households. But the partial contribution of input and TFP to the growth of output could easily be computed from the identity relationship of output and productivity growth. That is, growth rate in output is the sum of growth rate in inputs (index) and growth rate of TFP. So, growth rate of input is simply the residue after growth rate in TFP is deducted from the output growth rate.

On the other hand, the drastic fall in TCH (5.27%) for tef outweighed the significant growth in the TEC (3.82%) that ultimately led TFP to decline at a rate of 1.45% per annum. This suggests that technological regress narrowed the production possibility set and reduced the maximum attainable output using the given inputs. In other words, the regression in technology was far more powerful than progress in technical efficiency. Consequently, the growth in output of tef could be entirely ascribed to an increase in use of total inputs (particularly fertilizer). This implies that increasing use of fertilizer, if not complemented with improved seeds and other ways of better techniques, may replace only the amount of nutrient lost, with the result that tef yield will change only marginally or not at all.

Increasing resource use in tef production might be because of the incentive structure provided by the reform, particularly as regards the market and fertilizer supply (price of tef was most stable). The declining trend in tef productivity, however, could be due to two main factors: (i) tef production has been restricted to the most traditional farming system; neither adoptable improved varieties nor improved cultivation practices were available to rural farmers. The traditional varieties may therefore have lost their genetic potential and/or the old cultivation practices may not be compatible with the changing physical environment. If this is the case, one may expect similar problems in the production of many other crops for which no improved technologies have been developed. (ii) The other possible explanation is that the customary practices for tef production may aggravate soil erosion problems. It is believed that tef performs better on marginal soils than most other crops and hence it is usually used on such land. Since soil degradation in Ethiopia is increasing, there is a greater probability of tef being planted on marginal soils. In addition, from its very nature tef needs very delicately prepared seed-bed that in turn makes the land more vulnerable to run off and wind erosion. Such practices could worsen soil quality and lead to a further decline in productivity. Unfortunately, this aspect was not captured in our analytical model. It is therefore suggested that further research should be undertaken to elicit the causal factor.

Considering the analysis across the regions, although growth in outputs of all crops was generally positive, a disaggregated analysis of TFP growth by crops showed varied results. Changes in TFP for tef and wheat were not statistically significantly different across the locations under examination. TFP for tef was declining at a similar rate in both Ankober-Dinqi and Shashamane. Likewise, the improvement in TFP for wheat did not differ greatly in Debrebrehan and Shashamane. However, TFP for maize grew much faster in Shashamane than in Ankober-Dinq mainly because of relatively a higher growth in technological progress in the former location. Such a substantial difference may be due to differences in resource endowment and advantages related to proximity to the metropolis (for detail information see Table A4.9 in the Annex).

4.5.4 Conclusion

There is evidence that the economic reforms and institutional transformation induced use of more modern inputs at least in some crops. Nonetheless, its impact on changes in efficiency, technological progress and total factor productivity was wide-ranging. Thus, the performance of the crop production lines studied varied considerably. Only for wheat were both TEC and TCH positive. Hence, appreciable growth in TFP was recorded that accounted for the largest share of wheat output growth. On the other hand, the direction of change in technological progress and TE in the other two crops was opposite. In tef production, TE improved significantly, but the deterioration in its production technology was much greater, causing TFP to fall. Consequently, all the growth in tef output was derived from increasing use of inputs. Nevertheless, the contribution of human and oxen labor to tef production diminished over time. The changes in tef output can therefore be attributed entirely to an increase in the utilization of chemical fertilizer, as the area of land used for tef production was almost constant in the period. Conversely, technological progress was markedly advanced in maize production, but farmers became increasingly technically inefficient. TFP growth, however, was positive and contributed greatly to output growth. TE in maize declined because institutional changes and changes in the farmers' managerial skills did not take place in parallel with improvements in the maize technology.

Moreover, the result of the RTS indicates that particularly the tef and maize production lines were not scale efficient under the prevailing technology. While tef production experienced a decreasing RTS, maize production experienced an increasing RTS. This suggests the economic importance of reallocating resources (mainly human and oxen labor) to a greater extent towards production of maize and less towards tef in order to improve scale efficiency and enable the farmers to increase food grain supply.

Differences in agroecological potential did not significantly differentiate the TE and TFP of wheat farmers. Yet, its consequence on TE of tef and maize farmers across the locations was so high; but the gap reduced over the analyzed period. However, the difference in technological progress and TFP of maize farmers was amazingly increasing and the gap was growing.

Land is very limited, so an increase in labor will have only a limited effect, if any at all, on total production. In addition, the potential to boost production by improving technical efficiency was about to reach its peak as well (like in tef), given the traditional production system. In such a case improving TE without changing technological progress will therefore have a very limited impact. Therefore, the quickest solution for Ethiopian agriculture is to increase the use of modern inputs. Chemical fertilizer particularly is likely to have the greatest potential for increasing total production, because it was noted that fertilizer input per unit of land in Ethiopia is the lowest by all standards. In this case the efficiency of fertilizer market and its distribution to the rural farmers will be crucial.

5 Constraints to Technical Efficiency Improvement in Maize Producing Farmers

As noted in Chapter 4, contrary to tef and wheat, maize production efficiency tended to decline over the study period. This phenomenon attracted our attention and further analysis was undertaken to identify the main constraints hindering the technical efficiency of farmers in maize production.

5.1 Theoretical Concept of Causes of Production Inefficiency

There is a wide range of factors that can explain differences in technical efficiency. These range from physical, socio-economic circumstances to institutional environments. They can be broadly categorized as internal or external sources of inefficiency.

5.1.1 Internal Inefficiency Factors

The internal factors determining production efficiency are an individual's intra-governance, which includes the farmer's natural potential, education, experience and age. Such individual features crucially influence the ability of a farmer to use and manage the available resources efficiently, especially during new interventions. For instance, unlike the other crops, the improved maize seeds that the government has been introducing were largely *hybrids*. This type of seeds is technically not recommended for storage and reuse, but should be purchased from seed enterprises for every planting season. This may increase transaction costs (transport, information seeking and procurement, etc.) to farmers. Moreover, the agronomic practices for the improved maize production systems are quite different from farmers' traditional practices. For instance, row planting and spot-application of fertilizer are among the important components of maize production that are completely new. On the other hand, these are not included in the cropping systems for other crops, even with improved practices. Such practices may require experience and know-how. Consequently, the improved maize technology may only be adopted by a few farmers, thereby increasing the productivity gap among maize-producing farmers, with a resulting increase in technical inefficiency.

Transaction costs also play important role in determining production efficiency. This is the cost of all activities relating to gathering and processing information and solving possible disputes. Two major aspects of transaction costs are: (i) aspects associated with the organizational size of a firm, that is, cost of directing, negotiating with, organizing and monitoring workers in the firm (Fahlbeck, 1996). This constitutes the largest part of transaction costs but as farm organization of rural small farmers is generally limited to family management, the first aspect is not so important; and (ii) aspects related to economies of scale. This aspect of transaction costs may be worth considering, however, even though the rural farmers considered in this study are generally small. Although there is no harmonious position among economists as to what size (small or large) of a firm is economically optimal, returns to scale indicate the position of a farm's production. This optimum point is when the returns to scale are constant, which occurs at the minimum long-run average cost. However, it varies with the level of technological change, since it follows the relative importance of input use. It is noted that maize production was not operating at optimum economies of scale during the study period.

5.1.2 External Inefficiency Factors

External factors are the physical and institutional environments that affect farmers' incentives for production. First, some effects are due to the particular nature of the farms, e.g., soil and weather conditions. Hence, some of the farmers identified as inefficient by the present method may not really be inefficient if we could control for such individual effects. Development economists, however, very often pinpoint failure in market and non-market institutions as the major cause of firm inefficiency. Schmitt (1991) argued that policy failure and imperfections in factor and product markets restrict the extent and speed of resource use adjustments towards the efficient level. The presence of unreliable marketing channels and unstable prices for output in Ethiopia may increase risks and transactions costs. Another potential cause of farmer inefficiency could be poor organization and implementation of government development services, for example in technology diffusion/transfer processes (Assefa, 1995). The problem is more serious in the transition economies and is worse in the agricultural sector in general (Curtiss, 2002). Development of the financial market and enhancing its outreach to the rural economy is another crucial external factor affecting farmers' efficiency through its effect on use of modern inputs. In addition, poor infrastructure such as information networks and transport facilities increase transaction costs and are therefore also possible causes of technical inefficiency.

In the following section, we try to identify the most important factors affecting maize production efficiency in the study period. The mechanisms and the direction of their effects are elaborated at length.

5.2 Determining Farm-Specific and Institutional Factors Limiting Technical Efficiency

5.2.1 Technical Inefficiency Effects Model

More detailed modeling of the variables determining production efficiency makes the analysis more robust and more instructive for policy decision-making. A one-stage estimation procedure of the stochastic frontier production model was proposed by Battese and Coelli (1995). As they underlined, the one-stage model provides more efficient estimates than using the *two-stage*[25] estimation procedure. In the model specification, technical inefficiency effects are explicitly expressed as a function of a vector of firm-specific variables and random error and then integrated into the stochastic frontier model. Besides the specification given in (4.2), based on specification (4.1), the technical inefficiency effects, $u_{it,}$, is alternatively specified as:

$$u_{it} = \pi + \delta Z_{it} + \omega_{it} \quad ; \quad i = 1, 2, 3 \ldots N; \text{ and } t = 1, 2, 3 \ldots T. \tag{5.1}$$

Where u_{it} is the technical inefficiency effects assumed to be independent (but not identically) distributed as non-negative truncations of normal distribution with unknown constant variance (σ^2); and means ($\delta_o+\Sigma\delta_{kti}Z_{kit}$) that can be different for different firms and time periods. That is,

$u_{it} \sim N(\delta_o+\Sigma\delta_{kti}Z_{kit}, \sigma^2)$.

z_{it} is a (1×K) vector of firm-specific variables associated with the technical inefficiency effects of the *i*-th firm in time *t*. δ and π are an (k × 1) vector of unknown parameters to be estimated, and ω_{it} is an unobservable random non-negative variable to be i.i.d. with mean zero and variance σ^2_ω, i.e., $\omega_{it} \sim N(0, \sigma^2_\omega)$. It is obtained by truncation of the normal distribution at the point of truncation:

$-\delta Z_{it}$ (i.e., $\omega_{it} \geq -\delta Z_{it}$).

Technical inefficiency effects can be modeled in terms of various explanatory variables (external and internal factors), which could include functions of firm and management characteristics and period of observation.

---

[25] Two-stage estimation is the most widely applied procedure in investigating determinants of TE variation among firms, where firm-level efficiencies are first estimated and then regressed upon firm-specific variables (such as firm size, experience, management quality, market access, etc.). However, the statistical validity of this approach is debatable (for more detail, see Coelli, Rao and Battese, 1998).

Similar to the specification of u_{it} in equation (4.2), the hypothesis that technical inefficiency effects are not random is expressed by H_o: $\gamma = 0$. In addition, the null hypothesis that the technical inefficiency effects are not influenced by the level of the explanatory variables in equation (5.1) is expressed by H_o: $\delta'=0$, where δ' denotes the vector, δ, with the constant term, π, given that it is included in the expression, δz_{it}.

5.2.2 Variables and Hypothesis

Expected variables to explain technical inefficiency in maize production were selected based on results of previous studies in Ethiopia and other developing countries (see for instance Assefa, 1995; Mohammed et al., 2000). These variables were included in the model with modification of some and inclusion of other new ones (discussion below). For variables where we faced difficulties in obtaining quantified data, substitute or proxy variables were used. The mechanism by which the variables affect technical inefficiency of production could most often be interrelated. The hypothesized variables are discussed as follows:

District dummy: To capture the effect of agro-ecological differences on farmers' technical efficiency, a dummy variable is included. A more agriculturally suitable environment is expected to induce farmers to produce with a greater degree of technical efficiency than less favorable areas.

Financial situation of farmers: Availability of credit affects the ability of a farmer to obtain the necessary inputs at the right time and in suitable quantities. This, in turn, substantially affects the TE of a farm by affecting its use of modern inputs and the probability that it will use improved technologies. Credit for the purchase of consumption goods (mainly food) also influences TE through its effect on labor productivity. Thus, availability of credit is expected to be negatively related to technical inefficiency.

Extension visits: Extension agents provide farmers with new information on improved technologies, markets, etc. So farmers' contact with such agents can improve their access to improved inputs and farming practices, and hence their production efficiency.

Infrastructure: Distance to the main road, graveled or asphalted road, is used as a proxy for infrastructure such as access to markets, transport services and information (or generally transaction and transport costs). It is expected to negatively affect technical efficiency. The longer the distance from main roads the greater the transaction and transport costs are, and the poorer the access to information. Consequently, the TE of an individual farmer diminishes.

Livestock: The amount of livestock owned (representing physical capital) indicates the wealth of a farmer. Livestock provides working power, manure fertilizer and is a source of cash that can be used to purchase consumption goods and necessary farm inputs. In a rural economy, livestock can be used as collateral, and thus influences farmers' access to the credit required for purchasing modern inputs (Buchenrieder, 1996; Lelissa, 1998). In such a case, livestock have a complementary relationship with maize production in the study areas and may positively affect the TE of maize-producing farmers. On the other hand, production of livestock together with crops is a sort of diversification. According to the transaction costs argument, diversification has a negative effect on TE. This is because it is accompanied by higher production complexity, which could be associated with higher costs relating to management, monitoring and information acquisition. Therefore, the actual effect of livestock on the technical efficiency of crop production is an issue that would merit further research.

Farm size: There are arguments for and against large farm size in terms of the direction of its effect on farm efficiency. For the advantage of simplicity in management and transaction costs, small farm sizes are expected to be more efficient than large farms. However, many argue that large farms do have an advantage as regards the scale of operations and that they are more likely to employ modern practices and hence could be more efficient than small farms. So, as in the case of livestock, the actual direction of the effect of farm size is not clear. Since the land size allotted to each crop is generally small in a peasant farming system, total land size cultivated would better represent the size of farmers' operation, and hence better explain TE.

Age: Battese and Coelli (1995) suggest that the time variable in an efficiency model captures the benefit of experience and learning by doing in individual firms' efficiency. Here, the age of the household head was used as a proxy.

Consideration of three other, additional explanatory variables would make our approach different. These are: the educational status of a household (not only of the head), its dependency ratio, and farmers' perception of market prices. The direction of the effect of these variables was hypothesized as follows:

Education: In most of the studies hitherto undertaken, only the educational level of the household head was considered to affect efficiency of resource use, and reports often suggest that education does not have significant role in production efficiency (e.g., in Mohammed et al., 2000), while others show mixed results (e.g., Assefa, 1994; Bravo-Ureta and Pinheiro, 1997). In our analysis, however, we would like to appreciate the role of the educational level of all the other family members, including children and housewives, particularly where there is impetus for technological change. Educated members can better assist the household head in providing information and/or could have a better position in participating in decision-making relating to resource use. Because many farm operations ranging from planting, weeding and fertilizing through harvesting are performed by family members, their knowledge and understanding, particularly as regards the use of improved seeds and new agronomic practices, can be very influential. Furthermore, the number of educated household members also can represent a household's position in terms of human capital. Consideration of the educational level of all household members would therefore better explain efficiency in production than if one considered only the household head's education.

Dependency ratio: Large family size and a high dependency ratio are typical characteristics of many developing countries like Ethiopia.[26] Given that the labor market is poor and almost all farm work is done by family labor in the country, activities like taking care of children and older family members are expected to reduce the efficiency of farm production. Thus, the higher dependency ratio the lower TE is.

Farmers' perception of maize price: Farmers experience and anticipation of the output price is also hypothesized to affect farmers' production efficiency. Unstable and/or low output prices increase farmers' vulnerability to price risks that can greatly affect their resource use. In Ethiopia, after market liberalization the price of maize in particular tended to fluctuate more frequently, and in years of good harvest it drops drastically (discussed in detail in chapter 6). Therefore, the perception of the household head regarding the market price of maize was taken into account to represent farmers' reaction to price risks and its effect on their production efficiency.

5.2.3 Description of the Variables

As indicated in Table 5.1, 66% of the total sample maize growers were from Shashamane district, and of the total sample only 33% had access to credit services and 18% were able to contact extension agents two or more times in a year. Their educational level was low, as expected, with an average of 2 persons per household[27] being able to read and write, most of whom were schoolchildren. The sample households' dependency ratio is about 0.76 (76%), which indicates that one member should support him/herself plus about one dependant. Nearly all the dependents fall into lower age groups because of the

[26] According to the CSA (1999), the average dependency ratio of households in rural Ethiopia is estimated to be 87 percent.

[27] The average size of a household in sample areas during the study period was about 5.5.

short life expectancy of the rural poor. It is known that poor rural farmers are generally risk-averse. In our sample, about 56% of the sample considered the price of maize to be very risky or lower than the minimum level they expect.

Table 5.1: Description of the Hypothesized Variables

Variables	Description	Mean
District	1 if Shashamane, 0 otherwise	0.66
Credit availability	1 if a farmer has access to loan from any source, 0 otherwise	0.33
Extension visit	1 if two or more times in contact with extension agents in a year, 0 otherwise	0.18
Distance	Distance to a nearby asphalted/graveled road in walking minutes	40.00
Livestock size	Livestock size owned, converted into livestock units	3.57
Farm size	Total farm size owned in hectares	1.24
Family education	Total number of persons able to read and write in the household	2.05
Dependency ratio^{+}	Ratio of economically inactive to active labor force	0.76
Age	Age of the household head in years	43.00
Sex	Sex of the household head; 1 if male, 0 otherwise	0.87
Perception of price risk	1 if household head's perception is that maize price is low or risky, 0 otherwise	0.56

Source: Own description

Note: + Ratio of sum of members of a family under the age of 14 and over 65 to total members between the ages of 14 and 65.

5.2.4 Empirical Results

The maximum likelihood estimates of the CD frontier production function defined in equation (4.11) and technical efficiency effects (5.1) were carried out simultaneously. Unfortunately, the FRONTIER 4.1 software does not compute the marginal effects of each variable in the model. Thus, only the direction of the causal relationship of the hypothesized variables (not the magnitude of their effects) could be determined. Results of the estimation of the relationship between the TE of maize growers and farm household characteristics and institutional factors are shown in Table 5.2. The general sign and magnitude of the coefficients of the variables explaining TE in maize are credible. Many of the explanatory variables entered into the inefficiency model were individually significant in explaining the TE in the frontier model, and hence all the variables together cannot be rejected. The joint hypothesis that all non-intercept coefficients (that is, Ho: $\gamma = \delta_k = 0$ for all k) in the model are zero was thus rejected[28] by the data. This indicates that the joint effect of the inefficiency-explaining variables was significant for maize production. As indicated by the coefficient of γ, the value 0.51 implies that a substantial proportion of the total variability in maize production was associated with technical inefficiency. In other words, about 50% of the total deviation of actual output from the maximum attainable with best practice was ascribed to technical inefficiency.

With all other factors held constant, a negative sign of a parameter explaining inefficiency is interpreted as follows: an increase in the magnitude or frequency of that variable reduces technical inefficiency (or improves technical efficiency), and vice versa for the positive sign. The result shows that the efficiency explanatory variables for rural small farmers, such as credit availability, distance to the main road and number of livestock units owned statistically significantly influenced the technical efficiency of maize production. The three additionally hypothesized variables, namely number of educated family members, dependency ratio and households' perception of price risk, were also found to have significant impact. However, the other variables such as district dummy (consistent with the earlier finding), extension contacts, farm size and age of the household head had no statistically significant impact on the technical efficiency of maize-growing farmers in the study area in the period analyzed.

Credit availability showed an opposite relationship with technical inefficiency in maize production, implying that the better a farmer's access to credit is, the higher his/her technical efficiency is. Access to credit critically affects the use of modern inputs, intensity of use and timeliness. Many studies in Ethiopia (e.g., Beyene, 2002; Abebe and Mulat, 2003; Lelissa, 1998) show that the majority of farmers do not use improved seed, and most of those who use fertilizer apply it below the recommended rate because of financial constraints. These problems definitely affect the technical efficiency of farmers. In this regard, using the Tobit regression model, Mathijs, et al. (2001) show how transaction costs are eased through cooperative membership, facilitating credit and accessibility of inputs, and thus improving TE. Therefore, the availability of technology alone cannot be a universal remedy for production growth.

Family education has a negative sign, indicating that the larger the number of educated family members, the more efficient they become in maize production. More importantly, the interaction term of availability of credit and family education was highly significant (at 1%) in determining the efficiency level of maize production. Therefore, our analysis lends supports to Schultz's hypothesis regarding the effectiveness of education on efficiency in a modernizing environment. This is in line with the findings of Abdulai and

[28] The LR result favors the estimation of the frontier model over the OLS (Chapter 4), and is supported again here by the t-tests in the inefficiency model.

Huffman (2000) in Ghana. This clearly suggests the significance of combining both education and credit to improve TE in the production of maize.

Table 5.2: Maximum Likelihood Estimates of Determinants of Inefficiency in Maize Production

Variables Explaining Technical Inefficiency	Coefficients	t-ratio
Constant	3.193	4.014**
District	-0.064	-1.240
Credit availability	-0.212	-4.340**
Extension contact	-0.013	-0.120
Distance to main road	0.026	1.895**
Livestock units	-0.113	-3.251**
Average farm size	0.102	0.116
Family education	-0.085	-1.765*
Credit×Family education	-0.521	-7.861**
Dependency ratio	0.120	1.563*
Sex of household head	-0.432	-1.113
Age of household head	0.027	0.135
Age squared of household head	-0.002	0.031
Perception of price risk	0.241	2.672**
Sigma-squared (σ^2)	0.676	2.143**
Gamma (γ)	0.510	1.962**
Loglikelihood ratio function	-82.75	17.294

Source: Own computation

Notes: ** and * denote a 5% and a 10% level of significance respectively.

The dependency ratio of a household and the perception of the household head regarding the market price of the crop output were also found to have considerable impact on the TE of maize production. These explanatory variables have positive signs, indicating that the higher the dependency ratio, and the more a household head anticipates a low or fluctuating price, the higher the probability that the TE of the maize farmers will be lower. This calls for policy interventions to encourage and provide farm households with the necessary facilities to practice family planning. The importance of stable market prices for farm producers to revitalize an efficient maize production system has become imperative.

Holding livestock significantly reduced farmers' inefficiency. This finding is consistent with the research results of Assefa (1995) and Mohammed et al., (2000). Furthermore, the finding corroborates the investigation of Abebe and Mulat (2003), namely that use of improved technologies and hence incremental mean output was observed in the case of households with a better endowment position. The complementary role of livestock was therefore more important than its diversification effect as regards the TE of maize. Moreover, distance from the main road has a positive sign and is significant. This indicates that the farther a farmer is from the main road, the less access he has to markets or information, and hence the less efficient he becomes.

Contrary to the primary objective of extension service, farmers' contact with extension agents did not significantly influence their efficiency in maize production. In fact, this may not be surprising. From the field survey it was noted that the ratio of extension agents to farmers was unimaginably low; it was estimated at between 1:2,000 and 1:5,000. Moreover, extension agents were supposed to undertake other non-technical activities such as collecting taxes and government loans, which was actually an approach that was unacceptable to the farming communities. Under such circumstances, the insignificant role of extension agents in terms of the technical efficiency of maize production could be understandable. More astonishing are the findings of Mohammed et al. (2000) relating to barley production in south-eastern Ethiopia, and Seyoum et al. (1998) on maize production in eastern Ethiopia; they found that extension contacts in fact increased inefficiency.

In spite of the fact that maize farmers were operating at an increasing RTS (which is not efficient), farm size did not significantly affect their technical efficiency. A similar result is documented by Voigt et al. (2001) in Russian agriculture. There may be two possible reasons for this: either farm size is not a proper proxy for transaction costs, or the size of farm holdings of the sample farmers was generally small and there was little variability in this regard. The age and sex of the heads of the sample households also had no significant influence on the TE of maize in the study period.

5.3 Conclusion

Although technological progress in maize production improved in the study period, farmers' technical efficiency declined. The inefficiency analysis suggests that higher household education, access to credit, amount of livestock owned, distance from the main road (remoteness), unreliability of the market price and dependency ratio were among the major constraints inhibiting farmers from improving their TE, in conjunction with technological change. The study showed that the potential for increasing output through improving TE is very high. Particularly, increasing the availability of credit on a sustainable basis, enhancing education and reducing market risks through developing relevant institutions could help to curb the problem.

6 Market Liberalization, Price Instability and its Impacts on Modern Input Use and Food Production

In Chapters 4 and 5, we analyzed the (potential) growth in total factor productivity of the most important food crops and actual changes in technological and technical efficiency over a certain period of time after the economic reform, with relatively greater government support and better climatic conditions. We saw that increasing the use of modern inputs is crucial for boosting productivity and production. In particular, more intensive use of chemical fertilizer and improved seeds was found to make a significant contribution to enhancing productivity in wheat and maize production. Despite greater use of chemical fertilizer, it is also interesting to observe that the productivity of other crops (such as tef) deteriorated due to a lack of complementary improved technologies. Besides, it may not be ridiculous to investigate that technically efficiency of farmers producing maize declined regardless of improvement in technological progress in its production. One factor identified was that farmers were confronted with a high instability in market price for the product.

The analysis undertaken so far, however, was at the micro level, based on data from normal natural and economic situations and covering a short period. Other important factors, such as the general effect of price fluctuations and weather changes on farmers' practices were not captured in the previous chapters. The objective of this chapter, therefore, is to analyze the problem of market price instability on sustainable use of modern inputs and food supply at the macro level.

6.1 Introduction

Why, it is argued, does Ethiopia, given its potential for productivity growth and the major contribution of market liberalization to food security and agricultural growth, still continually face hunger? Severe food insecurity problems have been observed under almost all government regimes. During the study period, nearly half of the total population was vulnerable to transitory food insecurity, while some 5 to 7 million faced chronic food insecurity. High variability and shortfalls in domestic production are often cited as primary causes of food insecurity in the country. For instance, the dramatic growth of agricultural output in the couple of years leading up to 2001, the discouraging fall in product prices immediately thereafter in 2001/02 and the decline in fertilizer/improved seed use, followed by the miserable famine in 2002/03, were incidents that might be paradoxically difficult to understand.

The problem of food insecurity may be associated not only with production but also with marketing and distribution issues. Marketing policies affect the variability of food prices and hence supply, which ultimately affects the vulnerability of low-income households to food insecurity. In Ethiopia, since the inception of market liberalization, the government has withdrawn its market intervention, particularly in output marketing, trusting market forces to guide and stabilize the economy. The new role of the government since the reform is not clear in this respect; it has given little emphasis to reducing vulnerability to food crises. Measures to strengthen emergency capabilities and to develop an effective early warning system seem too weak, and efforts to establish strategic food reserves have been virtually absent.

The primary question may therefore be a problem of market failure: is it rational to assume that a market will perform well when it is basically undeveloped? Or are inputs and output markets functioning well in Ethiopia? More specifically, can the prevailing market institutions in the country encourage small farmers to use (more) modern inputs sustainably so as to enhance food production? The accumulated evidence does indeed indicate that market liberalization affects the rate of economic growth, and hence the extent to which households may be lifted out of poverty (Schuh, 2002). However, there is no guarantee that the free market system can work effectively in an economy where the market itself is not (well) developed. It is also argued that fully liberalizing prices could increase the risk of price instability, curbing the incentive to use improved technologies and ultimately worsening food insecurity problems.

Despite the considerable influence of price volatility on the performance of the agricultural sector in particular, little research has been done in this area in Ethiopia. The impact of price fluctuations on the use of modern inputs and on food supply remains largely unexplored. In this chapter, first of all, a brief review of the challenges for achieving food security in a liberalized market economy (in poor countries) is presented. Then, preliminary observations are addressed relating to price trends for major agricultural grain products and modern input use in Ethiopia. Last, the results of econometric analysis of the intensity of volatility in output prices and its impact on modern input use and food grain supply are discussed in detail, followed by some suggestions for policy measures.

6.2 Market Liberalization and Food Self-sufficiency: The Challenges

Within the broader concept of food security, two types of food insecurity are worth mentioning in terms of policy actions. These are "chronic" and "transitory" food insecurity. According to von Braun, et al. (1998), the former is a sign of poverty and is often caused by long-term failure to obtain adequate food. It requires fundamental changes in production technology and growth in the economy as a whole. The latter is caused by short-term fluctuations in production or in the price of food. In severe cases, the transitory type takes the form of famine and requires urgent and coordinated efforts, such as drawing on reserve stocks or urgent food aid/imports to withstand its shocks.

Zeller et al. (1997) have shown that access to credit is an effective mechanism for dealing with food insecurity problems. They emphasized that credit can provide capital for financing inputs for income generation to address the problem primarily of chronic food insecurity. In addition, they argue that other credit is also essential to stabilize consumption of food, and hence helps prevent transitory food insecurity problems.

For both micro and macroeconomic reasons, no country has ever sustained the process of rapid economic growth without first solving the problem of food security (Timmer, 1992; Delgado et al. 1994). At the micro level, inadequate and irregular access to food limits labor productivity and thus income per capita, and further reduces investment in human capital (Straus, 1986; Fogel, 1994). At the macro level, periodic food crises undermine political and economic stability, reducing both the level and the efficiency of investment (Timmer, 1989; Heidhues, et al 1992). The effect of food price instability on investment and growth throughout the entire economy is explicitly documented in the works of Timmer (1998):

> *"Instability in the food sector can have three macro-level effects. First, it can affect the quantity of investment through an increase in precautionary savings or a decrease caused by greater uncertainty. Consumers save to protect themselves against the effect of a possible increase in food prices, whereas producers (farmers) save to insure themselves against a sudden drop in crop prices. These precautionary savings are kept in liquid form, and might not contribute to economic growth. Secondly, it can decrease the quality or efficiency of investment (rate of return), because instability in food price gives little information relevant for long-run investments as it causes instability in all other prices in the economy. Finally, price instability can induce a bias towards speculative rather than productive investment activities and thereby slows down economic growth."*

Thus, food security can influence economic growth in two ways: through the impact on labor productivity, and through the quantity and efficiency of investment.

Consumers also benefit from stable food prices because they do not have to incur transaction costs arising from reallocating their budgets frequently, nor face the risk of a sudden and sharp deterioration in their real income. According to the findings of Timmer (1989), this benefit accrues disproportionately to the poor because they spend a larger portion of their budget on food and they are more sensitive to price changes than the rich. Note that the pain of price instability is highest for small scale subsistence farmers as they are producers as well as consumers, and given that they comprise the majority in the society the its consequence for the economy is enormous. Moreover, in poor countries, food price instability causes price instability in other sectors of the economy since the food sector dominates the economy.

An important reason for investing in a country's agricultural sector, therefore, is primarily to realize the potential to stabilize the domestic food supply and enhance food security. Therefore, policies to promote efficiency (like market liberalization) for rapid growth in agricultural incomes should be consistent with the broader perspective of food security. Two highly contrasting schools of thought have been arguing for a long time about these issues, and their debate is far from over. These are the free market ("pricist" or neoclassical economics) and stabilization schools.

6.2.1 The Free Market ("Pricist") Approach

According to the pricists, prices tend towards Marshallian equilibrium of supply and demand, and an "invisible hand" is assumed to guide and stabilize the economy. This approach advocates alignment of the domestic price to border prices (Schultz, 1978; Timmer, 1986). It is argued that agricultural prices should also reflect their opportunity cost, the border price. However, the international and domestic process that determines prices needs to be taken into account. As well as improving resource use efficiency, market liberalization is expected to reduce other social costs that emanate from government intervention in the market, such as rent-seeking, and hence improve the financial capacity of the government to finance research and infrastructure while domestic revenue is low and external support is not promising.

Despite the fact that self-adjusting agriculture does not really exist, the pricist approach seems to have won the attention of donors since the 1980s. Consequently, as explained in Chapter 3, liberalization of agricultural prices and markets is one of the main components of the reforms in the Structural Adjustment Programs of the present Ethiopian government. On the input side, the government explicitly abolished input subsidies of both kinds, i.e. explicit subsidies paid directly out of a country's fiscal resources, and implicit subsidies resulting from overvalued exchange rates that artificially make imports cheaper. On the output side, the reform encourages deregulation and reduction of government intervention in the marketing and distribution of products, e.g., removal of parastatals or minimizing their role.

Although the primary objective of price liberalization is to improve price incentives for farmers and to increase gains through more efficient resource allocation, many argue that such gains are generally found to be quite small in most analyses using a general equilibrium model (Narayana et al., 1991). Schultz (1978) also argues that the paradigm presupposes full information, perfect transport with minimum transaction costs, and a large number of both producers and consumers. These assumptions are far from the reality of many developing countries. Particularly for basic food products that have an important role in economic stability and for the welfare of the population, the free market paradigm is misleading. So, the gains from agricultural liberalization may be overstated.

Moreover, in a poorly developed market, farmers lack full information on future prices and hence perceive prices as unpredictable. In such a situation, liberalizing prices may exacerbate the instability of agricultural production, thereby increasing risk and uncertainty. The problem is therefore more serious in an economy where market information and transportation networks are not well developed (Eleni et al., 2002). In this regard, two kinds of price variability in agriculture are most frequently cited: year-to-year volatility in prices around the moving price level, and longer-term wide-swings in the price level. The former type results in the problem of uncertainty, which in turn adversely affects efficient allocation of resources by farmers. The latter leads to major changes in the income level of both consumers and producers in general.

The principal causes of price variability in agricultural products are high price inelasticities on both the demand and supply sides. On the demand side, consumers' income level, their consumption behavior and marketing costs significantly affect their response to price changes. On the supply side (in the short run), the major reasons are the immobility and/or the high cost of reallocating production resources, and the longer seasonal/perennial gestation period in agricultural products.

With the expansion of output due to technological changes in the face of relatively inelastic demand, output prices drop, resulting not only in adverse income consequences for technology adopters, but threatening the very process of sustainable technological advance itself. With falling prices and declining incomes, technological advance "sows the seeds of its own slow-down", that is, creates its own destruction. At best, early adopters benefit temporarily, but late adopters and non-adopters lose. In such cases, the social gains from technological advances accrue to consumers in the form of lower prices in the short run, resulting in an agricultural *treadmill*, which subsequently has a negative distributional effect on farmers. In the long run, this can have disincentive effects for modern input use that ultimately hinder food self-sufficiency (with a high cost to society as a whole) and the agricultural transformation process in general.

Jones and Beynon (1992) also argue that the ability of the private sector to respond to the liberalization process has been constrained by lack of credit, appropriate infrastructure, storage and transportation capacity, market information, and lack of the recurrent inputs. Based on Indian agriculture, Parikh (1992) has also come to conclude that agricultural trade should not go to world prices immediately. Domestic prices have to be such that they lead to a desired degree of self-sufficiency, particularly for goods that are protected by the rich countries. Finally, stability is desirable to protect the poor; given the variability in weather, domestic buffer stock operations in food grains have to be an integral part of agricultural policy (Narayana, Parikh and Srinivasan, 1991). In conclusion, a considerable body of literature indicates that the liberalization of input and output markets in the course of structural adjustment in the developing countries has been limited in its success. Let us turn our attention now to the arguments in favor of stabilization.

6.2.2 The Stabilization School

The stabilization school underlines the need for price intervention in the short run while aligning to long-run international parity. This approach is not merely about government intervention to control prices, but is based on genuine competitive intervention without significant budgetary loss. It assumes following the long-run price trends of international markets to enhance efficiency, while at the same time it argues that following short-run international prices tends to lead to efficiency loss. This approach encourages the development of competitive private marketing over time through the reduction of public intervention, as price stabilization becomes less important in the course of economic development (Ahmed, 1989).

Domestic prices have to be such that they lead to a desired degree of self-sufficiency. A reduction in the seasonal and regional price spread is generally advantageous. Stability is desirable to protect the poor, given the severe variability in food prices and supply. As Newbery and Stiglitz (1981) underline, stabilizing prices can bring immense benefit where farmers are very poor and highly risk-averse. Timmer (1989) and Brüntrup (1997) also point out that stable food prices enable households to increase investment in productive activities rather than holding stock.

The role of a certain level of government intervention in agricultural marketing is well founded in the history of agricultural transformation in the earlier period of western development and in the recent period of progress in the Asian economy. Stabilization policy has been successfully practiced in many countries in East and South-East Asia as well. They openly rejected the free market approach for primary staples and used favorable government intervention to support and stabilize agricultural prices (Eleni, 2002; Koning, 2002). In this regard, Indonesia's policy of stabilization by a parastatal agency throughout the 1970s and 1980s can be seen as a classic example of success in fostering intra-year and inter-year rice price stability (Ellis, 1993; Timmer, 1989). In the same period, Indonesia moved from being the world's largest importer of rice to self-sufficiency and then to net exporter in a matter of years. The main reason for its success in stabilizing prices were close monitoring of price trends and harvest predictions, relatively rapid responses to changing local conditions and interventions only at the margin of fluctuation with reliable and credible purchasing operations. Government policy simultaneously encouraged competitive private trading and the many small traders competitively participating took a 94% share of the market. At the same time, the government has been enhancing market efficiency by developing market infrastructures such as transport, information and communications.

Although the free market policy assumes that market forces dynamically adjust prices to equilibrium and stabilize the economy, in the Ethiopian the race between supply of and demand for agricultural products could rarely be equal. The first factor is the fallacy of "invisible hand" that assumes the prevalence of a large number of buyers and suppliers. In deed, in a subsistence economy, farmers are also consumers of large portion of their products. However, in Ethiopia only about 15% of the total population is living in urban or in non-agricultural sector. This number seems to too small to create sufficient demand for food products produced by the farmers, which constitute 85% of the total population, under good conditions. In deed, in a subsistence economy, farmers are also consumers of large portion of their products. However, farmers in agro ecologically poor potential areas do have little purchasing power from the open market, and hence totally depend on food aid. In this situation, it might be a dream to think of export as external demand while thousands of inhabitants have been still dying of hunger (FAO, 2002; WFP, 2002). Besides, the likelihood of the country's product to compete in the global market is gloomy.

Secondly, the forward and backward linkage of the domestic industries such as food processing industries and agriculture seem weak (see also the Granger Causality relation in Table A3.1 in the Annex). During the survey period, we could notice two primarily causes for the week interdependent between these sectors: (i) limited capacity of the local industries to absorb the agricultural products (many are still at infant stage) during a good harvest years, and (ii) partly low demand of the domestic food processing industries for local agricultural products mainly because of low quality of the products.

In the aftermath of structural adjustment, where the market is allowed to operate freely, complete government withdrawal may have deleterious consequences for an effective agriculture-based growth policy. Therefore, successful implementation of a completely free market in poor, sub-Saharan African countries like in Ethiopia may be illusory.

6.3 Trends in Prices, Modern Input Use and Food production: Preliminary Observations

In most studies relating to food security, food has been identified solely as cereals. This is because the share of cereals in total food consumption (measured in calorie equivalent) makes it the clearly dominant food staple in most developing countries of Asia, and most African countries. As noted earlier, wheat, maize and tef are particularly important in the food sector of the Ethiopian economy; they play a dominant role in the food market and hence have gained special priority in agricultural development strategy. The absolute variability (instability) in prices of these products has important policy implications for modern input use and adequate food production. This section focuses on analyzing trends in the output prices of these cereals vis-à-vis trends relating to the utilization of modern inputs. Descriptive statistics of some selected economic variables (1980 to 2003) are given in Table A6.1 in the Annex.

6.3.1 Trends in Prices of Modern Inputs and Cereal Products

During the socialist period, prices of fertilizer and improved seeds were subsidized and controlled, keeping them at a low level. Thus, there was only a slight upward movement. After the reform of 1992, the general trend in the prices of these inputs was always upward, with minimum fluctuations. Two main causes may be identified. The first cause was the increasing depreciation of the local currency, and the second was poor market performance (development) for such inputs. Mulat (1999) substantiated the latter problem. He established that the market for fertilizer was monopolized by a few regional government-affiliated importers and distributors, and participation by private firms in the market was negligible. Shortage of foreign currency, and hence dependency on donors for funds to import fertilizer, inhibited importers from purchasing it on the world market when the price is low. The market for improved seed was even less well developed, and thus it was unavailable altogether in the open market. Meanwhile, however, there was an increasing demand for these inputs, induced by the government technology extension program and the decline in productivity of traditional practices, particularly up to 2000. In the period from 1993 to 2001, the price of DAP rose from 176 to 265 birr per 100kg (a 50% increase), while the price of urea rose from 156 to 195 (a 25% increase). Similarly, prices of improved seeds increased dramatically in the same period. Improved maize seed (hybrid) increased from 178 to 235 birr per 100kg, improved wheat seed from 140 to 260 birr, and improved tef seed from 153 to 353 birr. These amounted to an increase of 32%, 86% and 131% respectively for improved seeds of maize hybrid, wheat and tef.

Conversely, prices of major agricultural products swing seasonally and were highly volatile over all the years after the reform. Both intra and inter-year price variations increased in the post-reform period. Seasonal ups and downs in producer prices were documented in Mulat (1999) and Wolday (2002). Prices fall dramatically immediately after harvest, from January to March. This is because nearly all farmers undertake the largest part (79%) of their annual sales at this time due to lack of appropriate storage facilities and the need to repay loans (particularly for the purchase of fertilizer and improved seed) and to meet other financial obligations (e.g., taxes). Hence prices generally drop sharply, which leads to a decline in the net return to the farm. On the other hand, food crop prices rise later (June to September) in the year when many farmers run out of stock and hence supply is relatively low, and some poor farmers even start to buy from the market.

Three of the most important cereals, namely maize, wheat and tef, all with average quality, were considered for illustration. Figure 6.1 depicts the yearly nominal price fluctuation for these crops in the capital, Addis Ababa. Except in 1994, from the time of market liberalization in 1991/92, prices rose continuously until the end of 1999 (although with high seasonal variability). Considering the period from 1992 to 1999, the nominal price of maize increased from 129 to 147 birr, tef from 207 to 264 birr and wheat from 157 to 209 birr per 100kg. This amounted to a rate of increase of 14%, 27% and 32% for maize, wheat and tef prices respectively. Yet, on average, this is much less than the rate of increase in prices of chemical fertilizer and improved seeds described above.

Following the substantial increase in total cereal production due to increased use of fertilizer and improved seed coupled with favorable rainfall immediately after the reform till 2001, in the subsequent years (2000 to 2002) prices frustratingly dropped sharply. Prices fell from their respective maxima in 1999 to minima of 45, 95 and 120 birr per 100kg for maize, wheat and tef, respectively. This amounted to a fall of 68%, 55% and 55% for maize, wheat and tef respectively. This is the largest recorded price slump in the history of Ethiopia. According to the theory of agricultural treadmill, all the benefits from use of improved technologies may run to the consumer, and producers may even lose, which ultimately can result in distribution effect. If such unforeseen circumstance also persists in the future, use of advanced technologies in agricultural production and the effort to transform agriculture may remain a serious policy challenge.

In general, periodic weather changes, lack of market information and the absence of storage facilities are mentioned among the main causes of the price instability problem. Farm households have no storage facilities, and the government has no buffer stock facilities to mitigate the instability problem. Even though the parastatal agency in Ethiopia claimed to stabilize prices, functioning competitively in the output market, its role was drastically reduced. It operated with little working capital and manpower, and thus made no significant impact on food security (Girma, 2002).

Table 6.1: Variability in Crop Prices and Output (measured in Standard of Deviation)

Observation period	Tef price	Wheat price	Maize price	Weighted average crop price
before 1991	22.67	20.82	12.81	18.14
1991-1993	42.83	31.95	32.65	35.55
1994-2000	29.33	45.24	31.28	34.75
2001-2003	3.74	49.10	57.61	35.52
1992-2003	36.95	44.69	46.33	40.32

Source: Own computation from CSA database

Variance is often used as a simple measure of volatility. The standard deviations of prices of the selected crops are presented in Table 6.1. The differences in the standard of deviation in the different periods constitute a significant structural break in the level of volatility of crop prices. On average, the standard deviation of individual and aggregate prices increased strongly, suggesting an increase in the volatility of the food grain prices after the reform. A similar situation could be observed with regard to their aggregate volume of production (Table 6.2).

6.3.2 Trends in Modern Input Use

Research work to improve the genetic potential of crops and soil productivity started in the country in the late 1960s. Use of improved seeds and chemical fertilizer, therefore, does not have a long history in Ethiopia. Until the early 1990s, the use of improved seed and fertilizer was largely limited to state farms and farmers operating on a cooperative basis, and to only a few smallholders located in the vicinity of research stations. Massive popularization and wider utilization of the inputs by small-scale farmers began after the reform in the early 1990s (Beyene, 2002), initiated by the Sasakawa Global 2000 extension project in 1993/94 and later taken over by the Ethiopian government. After the reform, small farmers used on average around 85% of the total chemical fertilizer and improved seeds used in the country, with the remaining 20% being used by state farmers and large-scale private commercial farmers. In Ethiopia, cereal food crops have been taking nearly 95% of total fertilizer used. Tef, wheat, maize, barley and sorghum are the major crops that receive the greatest proportion of both fertilizer and improved seeds consumed in the country, in descending order of importance (CSA, 2000; Shank, 1996). The first three major crops together consume over 75% of the modern inputs used in the Ethiopian agriculture.

Because of rapidly declining soil fertility and the limited possibilities of area expansion, the World Bank (1994) estimated that fertilizer consumption would have to grow at a compound annual rate of 17% between 1994 and 2000 in order to keep the food deficit at the level of 0.8 million tons. In Ethiopia, DAP and urea are probably the only chemical fertilizers used. Figure 6.2 portrays the general trends as regards purchased improved seed and chemical fertilizer consumption. It is clearly discernible that consumption of these modern inputs followed that of output prices (compare with Figure 6.1). Compared to the pre-reform period (before 1992), fertilizer consumption dramatically increased, with acceptable ups and downs. For instance, in 2001, about 2.6 times as much fertilizer as in 1992 was consumed by small-scale farmers, representing a 161% increase. Over the period between 1994 and 2001, fertilizer consumption grew at an average compound rate of only 5.61% (still far below the level recommended by the World Bank). According to Mulat (1999), this achievement was mainly owing to increased donor support in the form of funds for fertilizer imports, coupled with favorable climatic conditions in the stated period.

The intensity of fertilizer use is represented by the ratio of total volume of chemical fertilizer used in a year to total hectares of land under cereal food production. In general, it increased sharply over the last twenty years (1980-2000) at a compound annual rate of 17%. In the period between 1992 and 2001, the intensity of fertilizer use rose at the rate of 4.3%. For example, fertilizer use was about 52% more intensive in 2001 compared to 1992. It grew from 3.05 kg to 4.65 kg per hectare. In fact, this level is still tiny compared to even African average (20 kg per hectare) and incomparably much lower than international standards (95 kg per hectare. In most cases, fertilizer is used as a complementary input to improved seed.

Figure 6.1: Trends in Prices of Major Food Grains

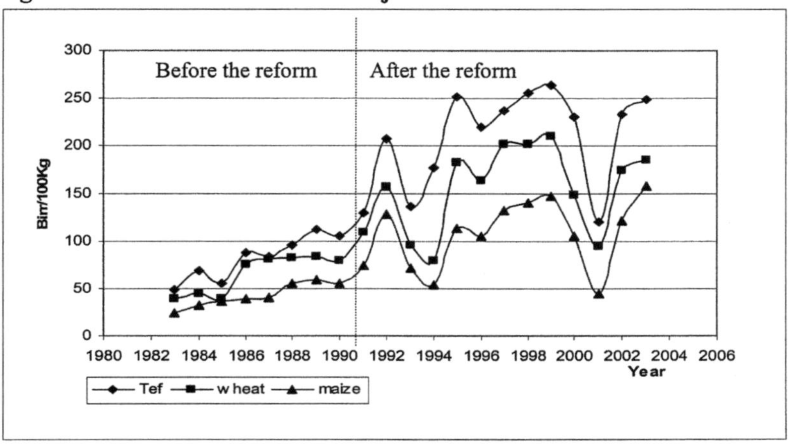

Source: Own compilation based on CSA database

Improved seed is almost always used in tandem with fertilizer. Similar to the case of fertilizer, use of improved seeds grew rapidly amounting to a compound growth rate of about 47% per annum after the reform (1992-2001). In 2001, for instance, about 2.6 times the amount of improved seeds was used compared to 1992 (i.e. an increase of over 160%). Hitherto, ESE (2001) revealed that overall utilization of improved seeds had only been 2% of the annual average national improved seed requirement, which is estimated at about 4 million tons for the different crops. The most important producers of improved seeds are the Ethiopian Seed Enterprise (ESE), agricultural research centers and multilateral firms (Pioneer) with the first of these usually producing over 90%.

Disappointingly, the use of both chemical fertilizer and improved seeds has fallen radically since 2001. Owing mainly to plummeting grain prices in recent years, farmers have reverted to the meager level they used 10 years ago; and because of the lag effect, input consumption did not rise along with the crop output prices in the subsequent years. Consumption of purchased improved seed fell much faster than fertilizer use. Such trends could have significant repercussions for future food production and food security.

Figure 6.2: Trends in Total Fertilizer and Improved Seed Consumption by Small Farmers

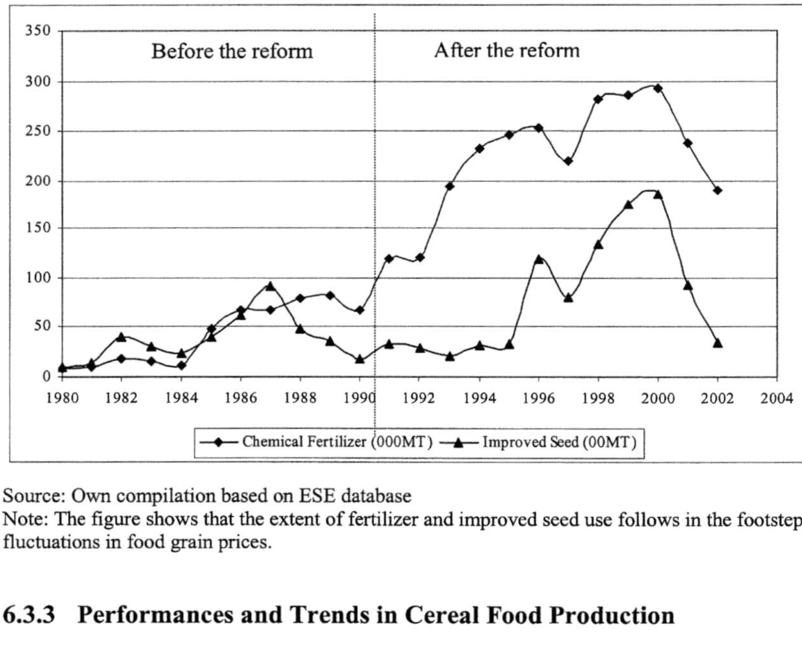

Source: Own compilation based on ESE database
Note: The figure shows that the extent of fertilizer and improved seed use follows in the footsteps of fluctuations in food grain prices.

6.3.3 Performances and Trends in Cereal Food Production

The overall trend of output of major food cereals before and after the economic policy reform is depicted in Figure 3.3 (Chapter 3). In this section, we discuss the growth performance of aggregate output of major cereal food products with regard to changes in area of land used and yield over the whole period after the reform.[29] Table 6.2 shows the growth situation of cereal production at national level in different periods after the policy reform. Considering the whole period

[29] Total growth in production can be derived as follows: $Y = A \times (Y/A)$; *and* $\log (Y) = \log (A) + \log (Y/A)$; where A is area in hectares, Y is physical output in tons, Y/A is yield (land productivity) and "log" is a natural logarithm. That is, the rate of growth in total production is equivalent to the sum of the rate of growth in the area of land used and growth in yield. Because yield is closely associated with the use of modern inputs, it does not necessarily represent the actual partial contribution of land to the total output. Nevertheless, it indicates the general direction of changes in the productivity of land and other inputs and their cumulative contribution to output growth.

after the reform (1992-2003), aggregate cereal output grew at an annual rate of 3.9%, a little faster than population growth. Because yield was stagnant or declining, all growth in aggregated production can be ascribed to an aggregate increase in the area for production and increasing use of fertilizer that some how maintained yield from further fall. Land expansion may be achieved through reallocating farmland to cereals and possibly by clearing communal forest or bush and extending cultivation to marginal areas usually left for livestock grazing. Aggregated yield of cereals remained stagnant even after the reform, despite the increased use of chemical fertilizer. The yield recorded was far below the international average, and even lower than the average yield of sub-Saharan countries (a global comparison is presented in Table A3.2 in the Annex). It ranges from 1.0 ton per hectare to a maximum of about 1.3 tons per hectare. This observation is consistent with the results we obtained for tef, with negative growth in TFP. This is because many cereals in Ethiopia continued to receive little chemical fertilizer and virtually no improved seeds, and there were no improvements as regards agronomic practices. So, even if farmers might improve their efficiency of production, absence of changes or deterioration in production technology and/or degradation of soil quality could lead yield (productivity) to stagnate.

Table 6.2: Average Annual Growth Rate and Variability of Cereal Output at Different Periods

Description	Annual growth rate (%)			Variability (SD)	
	1992-2003	1992-2001	1994-2001	Before 1991	1992-2003
Cereal output (000 MT)	3.90	7.90	6.50	719	1784
Cereal area (000 ha)	4.30	6.70	3.20	196	1254
Cereal yield (tons)	-0.14	1.20	3.30	0.13	0.12

Source: Own computation from the annual CSA database; SD is standard of deviation.

Over the period 1992-2001, aggregate cereal output increased spectacularly at a compound rate of 7.9% per annum. This resulted from a 6.7% annual average growth in the amount of land used for cereal cultivation and a 1.2% average growth in yield. During this period, the absolute increase in output reached 94%, of which 63% was due to an expansion in cereal area, while the remaining 27% was contributed by growth in yield. This was extremely high in the history of Ethiopian agriculture. The period 1994 to 2001 was particularly interesting for our study. As mentioned in section 4.3, this was a time when the government conducted considerable extension relating to improved technologies, and also a time when prices of products and weather condition were quite normal. In this specific period, cereal yield at national level increased radically at a compound rate of about 3.3% per annum, while the area of land cultivated grew at 3.2%, leading to output growth of about 6.5% per annum. This means that growth in yield contributed about 50% to the growth in total output of the major cereal crops, with the remaining 50% was due to expansion of the area under cultivation. Hence, total production showed mushrooming growth in this particular period; it more than doubled from its low level of 5,245 thousand tons in 1994 to about 9,578 thousand tons in 2001. This observation gives the impression that, under favorable climatic conditions and institutional support, there is considerable potential for increasing land productivity in such a way as to enhance food self-sufficiency. However, note that considering the whole period after the reform (1992 to 2003) growth in yield is close zero.

For the food security system to operate, the absolute magnitude of variability in output is critical. Production was not stable in Ethiopia, however. Output started to fall at the end of the observation year (in 2002/03), mainly because of a decrease in the use of chemical fertilizer and improved seed caused by drought and a drastic fall in output prices (detailed in the forthcoming sections). The last two columns of Table 6.2 show the degree of variability in production and its components as measured by standard of deviation (SD). In the period analyzed, the SD for aggregated cereal output and yield was 1784 thousand metric tons and 0.12 tons per hectares respectively. The coefficient of variation is 23% and 10% respectively. The output variability is much higher than in the pre reform, indicating severe instability in production. Production falls below average once in every four years.

Many factors are responsible for price volatility in agriculture. Supply lags in annual or perennial crops, weather changes, lack of developed infrastructure and poor market performance are among the major factors (for more detail, see Barkoulas et al., 1997; Breidt et al., 1998). Recurrent war and frequent policy changes and absence of good governance are also other important factors. During the research period, it was observed that the government repeatedly and unpredictably changed its policies, institutions and the working structures of experts. Thus, agricultural production was very vulnerable to many kinds of risks and uncertainties. Moreover, the dynamics of product prices have greater implications for food security in a liberalized market economy. Statistical descriptions of the variables used in the analysis are indicated in Table A6.1 in the Annex.

6.4 Econometric Analysis of Output Price Volatility and its Impacts on Modern Input Use and Food Grain Supply

In this section, we use econometric models on time-series data to measure the degree of volatility in farm product prices and its subsequent impact on the use of modern inputs and food supply. In order to avoid the problem of spurious regression in a time series, testing for the presence of a unit root is critically important. If a variable contains a unit root, then it is non-stationary and, unless it combines with other non-stationary series to form a stationary cointegration relationship, the regression involving the series can falsely imply the existence of a meaningful economic relationship (Harris, 1995). To estimate a regression model and obtain a reliable forecast, it is therefore essential that the data series is mean-reverting (i.e., a stationary process). Harris (1995) and Greene (1993) suggest that the Augmented Dickey-Fuller (ADF) and the Phillips-Perron (PP) tests are appropriate testing strategies to identify the order of integration of the time series (i.e., the number of times a variable needs to be differenced to make it stationary). The *Eviews* software performs both unit root tests, and the unit root procedure was applied for all data series used in the estimation of models 6.1 through 6.4.

During the study period, chemical fertilizers (DAP and urea) and improved seeds were the only modern inputs available to small farmers in Ethiopia. These inputs comprised farmers' biggest financial cost relating to agricultural production. As noted earlier, chemical fertilizers were most preferably used for tef, wheat and maize. Improved seeds were available almost exclusively for wheat and maize. For this analysis, data sets on annual quantities of fertilizer and improved seeds consumed by small farmers were obtained from the National Seed Enterprise and Fertilizer Agency. The data set of monthly prices of the cereals observed in the central market (Addis Ababa) was obtained from the annual reports of the CSA. In addition, the overall price indices for the major cereals and pulses were collected from the same source.

6.4.1 Estimating Volatility (Instability) in Grain Prices for Selected Cereals

6.4.1.1 The Model

Most statistical tools are designed to model the conditional mean of a random variable. Nevertheless, variance is often used to measure volatility, which is a key element in pricing theories. Therefore, modeling of variances of such important economic variables is an important development, particularly as regards the increased role played by uncertainty in the producer's decision-making process (Wooldridge, 2003). This is because price variability (volatility) is among the most important sources of uncertainty. Most studies assume that price variances follow persistent short-term stochastic models, such as the Autoregressive Conditional Heteroscedasticity (ARCH) model developed by Engel (1982) and/or the Generalized ARCH (GARCH) model developed by Bollerslev (1986). The models measure the degree of dynamics in conditional mean and variances. Studies using these models have found that the volatility process is highly persistent (Baillie et al., 1996; Bollerslev and Mikkelsen, 1996). The standard specification of the GARCH model is as follows:

$$p_t = \sum \gamma X_t + \varepsilon_t \text{; Mean equation} \tag{6.1}$$

$$\delta_t^2 = \omega + \alpha\varepsilon_{t-1}^2 + \beta\delta_{t-1}^2 \text{; Conditional variance equation} \tag{6.2}$$

where equation (6.1) represents the mean equation of the price of a commodity (p_t) as a function of exogenous variables (X_t) and an innovation term (ϵ_t) with a white noise (stationary) series. Equation (6.2) refers to the conditional variance; δ_t^2 is the one-period-ahead forecast variance based on past information; and ϵ^2_{t-1} (also called the ARCH term) denotes news information about volatility from the previous period, measured as the lag of the squared residual from the mean equation (6.1); δ^2_{t-1} (the GARCH term) is the last period's forecast variance. It is worth mentioning that model (6.2) is simply the result of the first model (6.1). The coefficients γ, ω, β and α are unknown parameters to be estimated. Nelson and Coa (1992) demonstrate that restricting $\omega > 0$, $\alpha \geq 0$, and $\beta \geq 0$ is a sufficient but not a necessary condition for the non-negativity of δ^2. Bollerslev and Ghysels (1996) suggest that the α coefficient can be interpreted as a measure of the short-run (immediate) impact of "news arrival" on volatility, while β controls the long-term evolution of the volatility process. The sum of α and β measures the degree of price volatility.

6.4.1.2 Analysis and Results

The ADF and PP test results for selected variables (crop prices) are presented in Table (A6.2 in the Annex). Both unit root test tools fail to reject the null hypothesis of a unit root in all price series at level, and hence they are all non-stationary. According to the PP test, we reject the unit root in the first differences, suggesting that the data series contains one unit root and is of integration order one I(1). Nevertheless, with the ADF test statistics, the variables were found to be of a I(2) process, which is in line with the findings of much of the literature which suggest prices are mostly I(2) (Wooldridge, 2003). Weighted prices and volume of production of cereals are also I(1) according to the unit root test results of both the ADF and the PP. Based on the PP test results, and according to the suggestion by Engle and Granger (1987), the variables under consideration are, by definition, cointegrated since they have the same order of integration. That is, the series are drifting together. So, their linear combination (a regression equation) at level can produce a long-run relationship between the variables.

The models presented in equations (6.1) and (6.2) above were simultaneously estimated using maximum likelihood. The dependent variable is product monthly price of the selected crops. Initially, many variables hypothesized to explain changes in cereal prices were included in the estimation.

These included: current and lagged output of major cereals (supply), money supply to GDP ratio (income), real exchange rate, a weather dummy (to capture productivity change and hence any shift in the supply function), and total volume of food imports (supply of substitutes). Moreover, prices of substitutes for each individual crop were included as determinants, while the price index of pulses was included to explain the overall price index of the major cereals.

It was found, however, that changes in prices of the selected crop grains were significantly explained only by their respective lag price, total cereal output and the weather dummy. All the other variables were statistically not significant, and the direction of the relation was economically not meaningful given the data and the model specified, and was therefore excluded from the model. (Further research could be suggested here.) As our main objective was to measure the extent of price volatility, only the results of model (6.2) of the GARCH (1, 1) analysis are reported in Table 6.3. In general, the results of the models are quite satisfactory. In each case, the test for normality of residuals was not rejected by the Jarque-Bera statistic, and all the other diagnostic tests moderately satisfied the classical regression axioms (not indicated here).

Table 6.3: GARCH (1, 1) Maximum Likelihood Estimation for Volatility of Major Food Grain Prices (1992-2003)

Coefficients	Tef price	Wheat price	Maize price	Cereals price
Mean (ω)	0.012	0.015*	0.023**	0.004
ARCH (α)	0.037	0.308**	0.436**	0.414*
GARCH (β)	0.544**	0.646**	0.532**	0.443*
($\alpha+\beta$)	0.581	0.954	0.968	0.857
Log likelihood	-15.21	-8.180	15.57	10.89
Durbin-Watson stat.	1.67	1.743	1.87	2.160
Akaike info. criterion	-0.610	0.743	-0.460	0.258
Schwartz criterion	-0.156	0.034	-0.159	0.021
F-statistics	7.56**	8.69**	21.14**	13.45**

Source: Own computation based on monthly data from CSA
Note: * (**) indicates a 5% (1 %) significance level

From visual examination of the data series (section 6.3.1), it was observed that, after market liberalization, prices of the major food crops became increasingly unstable over time. The results of the econometric analysis also reinforce the volatility in the prices of the selected series. Both the short-run (α) and the long-run (β) volatility measures are statistically highly significant, particularly for wheat and maize. Because we used monthly data, the long-run volatility index roughly represents seasonal and annual price variability. As may be clearly seen in the table, the β coefficients are much larger than the α coefficients in all price series, implying that seasonal and annual variability in prices are more serious than monthly changes. For these two crops, the sum of coefficients ($\alpha+\beta$) is close to unity, implying that volatility shocks are quite persistent (Baillie and Bollerslev, 1989). It is worth mentioning that the latter two crops are those for which improved varieties and fertilizer are predominantly used. The volatility in the aggregated prices of major cereals is also appreciably significant at the 10% level. The result is in line with the observations of Wolday (2002). The implications of such food grain price volatility were clearly manifested in the degree of modern input use and food supply. The next section discusses the empirical analysis of the consequences of such price volatility for modern input use and major food crop production.

6.4.2 Estimating the Impact of Price Instability on Modern Input Use and Food Grain Supply

From Figures 6.1 and 6.2 discussed above, it is evident that the trends in quantities of modern inputs used and prices of cereal outputs are drifting together. Such a waving trend indicates the existence of a cointegration relationship in the data series considered. To empirically establish cointegration, we first performed the conventional ADF and PP tests for all series at their levels. The unit root characteristics of the variable series are presented in Table A6.2 in the Annex. According to the result obtained, we cannot reject the null hypothesis of a unit root at any significance level. This outcome implies that the variables are of a higher level of order of integration. Further tests were then performed on the first and second differences. In this case, the application of the ADF and PP tests gave us somewhat different results. According to the PP test, all the variables under investigation were found to be a non-stationary first order process I(1) at least at 5% level of significance (i.e., the first differences are stationary I(0)). But with the ADF test, the results are mixed, some variables are I(1) and some others are I(2).

It is clear now that estimating a regression equation at the level of the variables is problematic (because of non-stationary problem), and differencing the variable (though stationary) is not a solution since this removes any information about the long run. A more suitable approach is to convert the dynamic model into an error correction formulation, because it contains information on both the short-run and long-run relationships of economic variables (Harris, 1995). This is possible if variables are cointegrated, which we need to test first.

6.4.2.1 Cointegration Tests

The technique of cointegration is widely used as a means of identifying and avoiding spurious regression. Most notably, it provides tools for the application of a dynamic error correction model that accounts explicitly for the dynamics of short-run adjustment towards long-run equilibrium. Economic variables are said to be integrated if their linear combination is stationary even though the individual variables are non-stationary (Engle and Granger, 1987; Harris, 1995). In our case, each series has been investigated for the presence of a unit root. Thus, before we run a regression analysis we first need to test whether the variables are cointegrated and identify the rank of cointegration over the sample period.

Several alternative approaches have been proposed to test for cointegration of series. Of these, the most popular are the approach put forward by Engle and Granger (1987) and that of Johansen (1988). The first of these is most commonly used to test for the existence of one cointegration relationship in a single equation. It depends entirely on the outcome of the unit root tests, and assumes that a cointegration relation exists if the variables under investigation are integrated of the same order, and if the residual of their linear combination has a lower order of integration varying randomly around some fixed level (Greene, 1993).

The Johansen approach, however, argues that the unit root tests are not reliable for establishing a cointegration relationship between variables; they often suffer from poor size and power properties (i.e., they tend to over-reject the null hypothesis of non-stationary when it is true and under-reject the null hypothesis when it is false). Harris (1995) further argues that having the same order of integration for all the variables under consideration is not a necessary condition (unless the variables n=2) for cointegration to exist. In practical application, cointegration can exist when there is a mix of I(0), I(1) and I(2) variables in a multivariate model. This corresponds to our findings in the unit root tests discussed above. Given that we are working with multiple variables, the Johansen approach is most appropriate. The *Eviews* software implements a Vector of Autoregression (VAR)-based cointegration test using the methodology developed by Johansen (1991; 1995).

6.4.2.2 Variables Used in the Model

In a liberalized market economy, the demand for (modern) inputs like chemical fertilizer and improved seed depends on expected farm returns, which in turn are partly dependent on the expected price of the product and the expected quantity of output. The quantity of output itself is determined by the productivity and intensity of modern inputs used. Moreover, the volume of production itself can affect market surplus, and thereby output prices. So, cereal output, cereal prices and modern input use are mutually interrelated. Specifically, the following variables were considered in the analysis:[30]

[30] The volume of food imports (purchased and aid) was expected to depress local prices and hence modern input use; the money supply to real GDP ratio (representing income) was

Dependent variables:
MDI: annual quantities of modern inputs used measured as weighted volume of chemical fertilizer and improved seed used (in tons);
CY: annual total volume of output of major cereal food crops (in tons).
Explanatory variables:
CA: annual area of land cultivated for the major cereals (in ha);
WDR: weather dummy variable (1 bad, 0 otherwise). One year lag dummy is considered since the main rain season in Ethiopia is from June to September for which its effect is harvested in following year from January to March.
PCSD: annual conditional standard deviation of prices of major cereals (representing volatility and obtained from estimation of model 6.2); and
RPR: annual relative price ratio (price of cereal output to price of modern inputs).
All variables except WDR and PCSD are in a natural logarithm.

We initially assumed that volume of output of cereals (CY), level of modern input used (MDI) and relative prices ratio (RPR) are endogenously interdependent. However, before embarking on the regression analysis, thorough identification of the direction of the causal relationship of the vectors of cointegrating variables is imperative. Given the data for the selected variables in the specified period, the results of a pairwise Granger causality test are presented in Table A6.3 in the Annex. The results reinforced a two-way relationship between series of CY and MDI at the 5% significance level. This implies that there is mutual interdependence (two-way causation) between the cereal output to be produced and the level of modern input use. More technically, the quantities of modern input used and food grain supply are cointegrating endogenous variables. On the other hand, the hypothesis that cereal output or intensity of modern inputs used does not "Granger-cause" output-input price ratio was not rejected at the 10% significance level, but the converse was rejected. Consequently, the RPR was considered as a weakly exogenous variable in the model estimation. Thus, number of endogenous variables is reduced to p=2, CY and MDI.

Before proceeding to the cointegration test, the appropriate lag length was first chosen using the loglikelihood ratio (LR) on results of VAR estimates, each

---

expected to positively affect food prices and hence modern input use; and export earnings were hypothesized to influence the supply of modern inputs directly through their effect on the financial capacity of importers, and hence increase the use of inputs and food grain supply. However, none of these variables was found to be statistically significant. This may be because food aid was directly targeted the needy poor people; and food imports were not also strong enough to affect cereal prices. Moreover, the government depends to a large extent on multilateral and bilateral support for fertilizer imports. Thus, export earnings have little effect on fertilizer imports. The effect of growth in food processing industries was found to be unimportant for the reasons mentioned above, hence exclude in the report. The absence of a significant relationship between cereal prices or modern input use and money supply to real GDP ratio is not clear. Admittedly, credit supply and extension services to farmers could have a potential effect on modern input use. But time series data on these variables were not available, and hence they were not considered in the analysis.

with different lags (details of the procedure is given in *Eviews 3 User's Guide in the Help menu*). In addition, we assumed that the data series contains a linear deterministic trend and intercept, but the cointegration equation includes only a constant. The number of cointegrating vectors (rank of cointegration vector) was determined on the basis of the maximum eigenvalues. Accordingly, the test results based on the maximum eigenvalues and the LR test statistics indicate two cointegration relations at the 5% significance level (Table 6.4). The estimated cointegration equation gives us an estimate of a long-run relation between variables.

Table 6.4: Johansen Cointegration Test (1980-2003)

Series: MDI CY
Exogenous series: CA WDR PCSD RPR

Eigenvalue	Likelihood ratio	5% Critical value	1% Critical value	Hypothesized no of CE(s)
0.6834	26.567	15.41	20.04	None **
0.2196	4.712	3.76	6.65	At most 1 *

Loglikelihood 27.204
Normalized cointegrating coefficients: CY - 0.05758 ×MDI - 8.091

Source: Own computation
Note: *(**) denotes rejection of the hypothesis at the 5% (1%) significance level.
The LR test indicates 2 cointegrating equation(s) at the 5% significance level, assuming lags interval 1 to 2.

6.4.2.3 The Model and Estimation

A single-equation approach is applicable only when there is a single unique cointegration vector and when all the right-hand-side variables are weakly exogenous. From the results of the Johansen cointegration and Granger causality tests, we have ascertained that the variables under consideration have a cointegration relationship and that some variables are endogenous. In such a case, a single-equation approach is inefficient (it does not lead to the smallest variance). For this reason, a Vector Error Correction model (VECM) has been particularly advocated as a way to estimate dynamic relationships among jointly endogenous variables; by using this model, the Ordinary Least Square regression becomes an efficient way of estimating each equation in the system comprising a common set of (lagged and predetermined) regressors. Unlike the Vector of Autoregression model, the VECM assumes the presence of cointegration (long-run relationships) and imposes cointegration restrictions among the variables by including error correction terms for the cointegration vectors. Thus, the VECM restricts the long-run behavior of endogenous variables to converge to their cointegrating relationships while allowing a wide range of short-run dynamics. Long-run relationship is the manner in which variables drift together, and the short-run dynamics is the relation between the deviations of a dependent variable from its long-run trend and deviations of independent variables from their respective long-run trends.

Supposing Y_{t1} and Y_{t2} are two endogenous variables, then the VECM can be formulated as follows:

$$\begin{bmatrix} \Delta Y_{1t} \\ \Delta Y_{2t} \end{bmatrix} = \gamma \begin{bmatrix} \Delta Y_{1t-1} \\ \Delta Y_{2t-1} \end{bmatrix} + \begin{bmatrix} \alpha_{11}\alpha_{12} \\ \alpha_{21}\alpha_{21} \end{bmatrix} \begin{bmatrix} \beta_{11}\beta_{12} \\ \beta_{21}\beta_{22} \end{bmatrix} \begin{bmatrix} Y_{1t-1} \\ Y_{2t-1} \end{bmatrix} \quad (6.3)$$

If the coefficients of the second term on the right-hand side of the equation are represented by Π and the non-stationary but cointegrating endogenous variables by Z, and considering the inclusion of weakly exogenous explanatory variables (X_t) and the vector of innovation ϵ_t, the VECM can be rewritten as

$$\Delta Z_t = \Gamma_1 \Delta Z_{t-1} + \ldots + \Gamma_{k-1} \Delta Z_{t-k+1} + \Pi Z_{t-1} + \Psi X_t + \varepsilon_t \quad (6.4)$$

The parameter Π is an error correction term and can be factorized into $\Pi = \alpha\beta`$, where α represents the speed of adjustment to disequilibrium and β is a matrix of long-run coefficients. The model (6.4) therefore provides information on short and long-run adjustment to changes in the Z_t, via the estimates of Γ_i and Π respectively. Thus, the subsequent impact of price instability, including other variables, on the use of improved technologies and food grain supply was estimated by applying the VECM.

6.4.2.4 Results

Initially, all variables hypothesized to affect modern input use and food cereal grain production were included in the model estimation. Then, only important variables such as changes in the relative real price of cereals to the price of improved seed and fertilizer (ΔRPR_{t-1}), conditional standard deviation (volatility) of the major cereal prices ($PCSD_{t-1}$), changes in area used for cereal (ΔCA) and the weather dummy (DWR), were chosen as determinants of demand for modern inputs (MDI) and food grain supply (CY). Lagged series of changes in volume of modern input used (ΔMDI_{t-1}), and lagged cereal output (ΔLCY_{t-1})[31] are conventionally included in the estimation of the model. Variables ΔRPR_{t-1} and $PCSD_{t-1}$ are seemingly correlated but are not significant at 5%; we therefore have no fear of encountering a multicollinearity problem.

[31] Only a one year lag of output produced was considered in the analysis as two or more lags were not meaningful. This could be because agricultural production by small farmers in Ethiopia is only hand to mouth and so long time output accumulated that can affect future production decision is limited.

Estimation of the VECM has provided quite satisfactory results. Table 6.5 summarizes the results of the analysis. The current level of modern inputs used and the volume of cereal food crops produced were slightly affected by a one-year lag of most of the variables included in the model except the area of land cultivated. This suggests that past experience/information is of immense importance for farmers' decision-making in current production. This could be because information on future prices or forecasting future market conditions was lacking due to the poor development of the market and market information. All variables included in the model have the expected signs and are statistically significant, except lagged cereal output (ΔCY_{t-1}) on the prevailing grain output supply (ΔCY_t). The error correction terms (EC_{t-1}) are statistically significant for both quantities of cereal output and modern input used. This suggests that deviation in modern input use needs about 3.5 years (28% per annum) to be restored to its equilibrium level, and output supply takes about 2 years (50% per annum). This can be considered a long time.

The positive coefficient of ΔMDI_{t-1}, though small, for modern input use could have been caused by the effect of farmers' experience with the inputs. Moreover, a change in previous output (ΔCY_{1-t}) has a positive and significant effect on the use of modern inputs in the consecutive year, as they greatly affect farm income and, therefore, the capacity of farmers to purchase inputs. The impact of modern input use (ΔMDI_{t-1}) on output supply of the major crops considered was positive and statistically significant at the 1% level. The impact of change in area of land cultivated for cereal (ΔCA) on food supply was significant but not in terms of use of the modern inputs. This may be partly because of the limited capacity of farmers to purchase modern inputs in quantities proportional to the amount of land cultivated. Moreover, at national level, the area cultivated had a much stronger impact on food supply (as explained by its larger coefficient) than modern inputs in the pervious year. This suggests that cereal production is heavily dependent on traditional resources, although this is clearly not sustainable.

The lag weather dummy is also statistically significant. As discussed earlier, crop production in Ethiopia is almost entirely rain-fed, and irrigation is scarcely practiced. Thus, agricultural production is very vulnerable to weather changes. More importantly, the impacts of the change in the relative price ratio (ΔRPR_{t-1}) and cereal price volatility ($PCSD_{t-1}$) on both modern input use and cereal grain supply are incredibly high, with a much greater effect on the former. Interestingly, as indicated by their coefficients, the positive effect of the relative price on the use of modern inputs, and hence increased food grain supply, is offset by the larger negative effect of the high grain price volatility. Output price instability, therefore, is unambiguously halting the incentive to use modern inputs and ultimately worsening food insecurity problems and blocking the road to agricultural transformation.

The shocks in prices and rainfall have a negative multiplier effect at sectoral level in subsequent years. That is, the shock in one period spills over into the next. For instance, drought or a sudden price fall in one year deprives peasants not only of current income but also weakens their capacity to buy improved seeds and fertilizer for the coming production season (as they have no opportunity to access the financial sector). If the incidence of the problem becomes severe, it could also lead to the sale or loss of wealth (selling and/or losing assets, in particular oxen) to pay their debts. This problem was observed in Ethiopia in 2001/02.

Table 6.5: Estimates of VECM for Modern Input Use and Food Grain Supply

Variables	Change in modern input use (ΔMDI_t)	Change in cereal output supply (ΔCY_t)
EC_{t-1}	-0.283^{**}	-0.508^{*}
ΔMDI_{t-1}	0.013^{*}	0.033^{**}
ΔCY_{t-1}	0.186^{*}	0.132
ΔRPR_{t-1}	0.214^{**}	0.145^{*}
$PCSD_{t-1}$	-0.252^{**}	-0.126^{*}
ΔCA	0.023	0.322^{**}
DWR	-0.491^{**}	-0.183^{*}
Adjusted R^2	0.761	0.594
F-statistics	21.64	16.18
Loglikelihood	-232.88	-207.55
Akaike AIC	-0.895	1.32
Schwarz SC	-1.45	-0.748

Source: Own computation

Note: Asterisk *(**) denotes a 5% (1%) significance level.

6.5 Conclusion

Liberalization of output markets in the course of structural adjustment in Ethiopia has merely limited success of agricultural transformation. From the time of market liberalization till the study period, both prices and volume of production of major food crops have become much more volatile than ever before. The results of the empirical analysis showed that the instability in the prices, coupled with recurrent weather changes, had a significant discouraging effect in terms of use of modern inputs by the small farmers and subsequently severely affected food grain supply. Overwhelmingly, farmers were trapped by twin problems, very low product prices in good harvest years and very low yield in bad years, both effects leading to poor farm returns and hence persistent food insecurity problems. On the one side, intensification of modern input use could not provide farmers with any guarantees, as prices of their products fell in successive years (given normal weather conditions) to the extent that they were unable to pay back they debts they had incurred to purchase the inputs. This is mainly because of inadequate effective demand in the economy. On the other side, low levels of modern input use (plus bad weather) reduced yield and deprived farmers of adequate output and hence low farm return. Furthermore, the poor farmers, as net purchasers of food, were confronted with high food prices, and thus became increasingly vulnerable to hunger.

Consequently, the government became increasingly dependant on food aid, and encouraged private traders to import food items in an effort to fill the gap created by the shortfalls in food production. However, purchasing food imports can drain foreign exchange and weaken the country's capacity for importing capital goods that could be more useful for further growth. Moreover, a sudden withdrawal or curtailment in the volume of food aid, or poorly timed delivery, has immeasurable social costs. Given that foreign aid is not a genuine option from the socio-economic point of view and imports of food cannot adequately fill the food gap due to lack of foreign exchange, and as long as there is the potential to produce enough food locally, the government should revise its agricultural development policy to secure the food requirement of the population and enhance overall economic growth.

Genuine government intervention to stabilize the food grain system, at least in the short run, until the domestic market is sufficiently developed, should therefore not be considered inconsistent with market liberalization. Small farmers have to be protected from price risks until their management skill is well developed to be able to properly operate in a free market system, and until they economically become better off to withstand both external and internal shocks. Development of market information, storage facilities and increased use of irrigation could help to curb the problem in the long run.

7 Summary, Conclusions and Policy Implications

7.1 Summary and Conclusions

Regardless of many controversies concerning agriculture, the literature provides ample evidence that both developed countries and contemporarily fast-growing countries have benefited from the immense role of agriculture in the process of their economic growth and structural transformation. A comparison of fast-growing Asian countries with sub-Saharan African countries showed that an increase in agricultural income, whether from resource expansion or productivity-enhancing technologies, is the most important factor that explains differences in the evolution of their economic performance. Failure to give due emphasis to agriculture first of all impedes the food production system and undermines food security; it hinders the transfer of resources necessary for the expansion of non-agricultural sectors, and ultimately slows or even prevents the process of structural transformation. We ascertained that both efficient prices and technological innovation in agricultural production are integral parts of successful development processes, provided they are suited to the unique social, institutional, and natural circumstances of the less developed countries.

In Ethiopia, three quite different political regimes, each with its own distinct development policies and strategies, have been experienced in the past five decades. While the Feudal-bourgeois regime practiced both outward and inward-oriented development policies, the Socialist regime was strictly inward-oriented. In both of these earlier regimes, the primary focus was on the development of industry, with little attention being paid to agriculture, the dominant sector of the economy. Coupled with periodic drought and civil war, the unfortunate economic policies of the earlier regimes severely damaged all sectors and most seriously agriculture. Eventually, the dismal performance of the Ethiopian economy convinced the present government of the necessity of sweeping policy reforms and institutional changes to replace the previously centrally planned economy with a market oriented economic system. Accordingly, since the early 1990s, the government has undertaken a process of reform, involving primarily market liberalization, privatization, promotion of the financial sector and agricultural research and extension institutions.

The growth of the agricultural sector in Ethiopia is undoubtedly important, as it constitutes the bulk of the economy in terms of both human and material resources. Hence, agriculture is required to show dynamism and growth if it is to alleviate poverty and to promote growth of the economy as a whole. The current government has designed development policies and strategies in such a way that agriculture receives the highest priority and plays the leading role in overall economic development. Accordingly, a rural based and export-led economic development strategy termed *Agricultural Development-Led-Industrialization* (ADLI) has been formulated. The strategy centers primarily on improving the productivity of smallholders to increase food supply and secure the food requirement, and then to boost surplus for industrial growth. It also encourages the industries to rely more on use of domestic agricultural surpluses. At the heart of the strategy lies the crucial importance of improving resource use efficiency and technological progress. Therefore, the radical changes accompanying the economic and institutional transition in Ethiopia are aimed at initiating sustainable economic growth to improve the living standard of society in general, and primarily to boost agricultural productivity and food production in particular.

However, agriculture's potential as an engine of growth and development and its potential for meeting the food demand of society hinges upon the performance of markets and price stability to guide efficient resource use. Because of institutional weaknesses or market failures that can depress the efficiency of economic agents, we still perceive a significant gap between expected and actual efficiency and growth levels. Furthermore, initiating and managing the take-off for productivity growth through use of improved technological innovation is perhaps the most difficult task awaiting developing countries in general, and especially those countries which, like Ethiopia, have experienced persistent stagnation or even decline in their economies. In addition, the agro-ecological diversity of the country can significantly influence growth in agriculture through its effects on choice of and access to technologies and infrastructure. Consequently, agriculture may be less responsive to policy changes in Ethiopia and it may even be inflexible when confronted with an unstable political environment and bad governance. This study set out to assess the impact of policy reform and institutional changes on the performance of agriculture as a whole.

The study began by briefly reviewing the overall macroeconomic performance of the country in the post-reform period compared to the pre-reform era, and assessing the position of agriculture in the economy. The government development expenditure portfolio showed a spectacular increase for agriculture, education and road construction, but, somewhat surprisingly, a sharp fall in expenditure relating to the industrial sector. With regard to industry, the government may have assumed that the private sector would fully participate in and promote the industrial sector without delay after the reform. Contrary to the earlier regimes, the government may have underestimated the contribution of this sector while emphasizing the role of agriculture in the economy. However, given that the process of privatization has been slow to get under way, and the legal system concerning property rights is still developing, private sector industrialization has been delayed, with the consequence that this heavily one-sided development strategy may prove infeasible.

A simple log linear growth model was applied to estimate the trend and growth of the sectors and the overall economy. While slight positive trends in the growth of all sectors could be observed, no appreciable change was achieved in the growth of real gross domestic product and food production per capita;

they remained stagnant. Nonetheless, external debt was growing much faster than ever before. Although no sizeable change was observed in the share of industry in the economy, the share of the service sector was growing quickly and it was rapidly becoming the largest (>60%) contributor to growth of the economy. While the medium and large industries were growing significantly, the overall growth of the industrial sector disappointingly, however, remained stagnant due to slow growth in the small-scale and cottage industries. Its share in the economy also remained constant at about 10%, whilst its relative contribution to economic growth plummeted by half (from 25% to 12%) after the reform. Although the share of agriculture fell by about 10% after the reform (in line with conventional theory of development), the role of agriculture in the growth of the Ethiopian economy was below our expectations; it contributed only 17% to economic growth after the reform compared to 9% before the reform. In addition, the share of the non-agricultural labor force only grew from 10% to 15% over the past twenty years, and its growth rate in fact fell after the reform, mainly owing to the increase in the capital intensity of the industrial sector (contrary to the ADLI's strategy). Hence, the vast majority of the labor force remained in agriculture. In general, this implies that it will take a considerable number of years for structural transformation to occur.

Moreover, even with an increase in the volume of agricultural exports (but with little diversity), the trade and current account balances in fact deteriorated after the reform. This was because of a much larger increase in imports of goods (particularly consumption goods) than exports, and partly because of a declining trend in world prices for primary goods (particularly for coffee, the country's main export). The other concern for the economy is that both agriculture and the domestic manufacturing industry were increasingly dependent on imported raw materials, while the linkage between the two sectors was weakening. These circumstances are likely to be unsustainable from the theory of development perspective. More detailed analysis on agricultural performance and constraints was carried out at micro and macro levels.

At the micro household level, this study attempted to analyze the extent by which technical efficiency (TE), technology (TCH), total factor productivity (TFP) and intensity of input use were changing and to estimate their relative contributions to output growth. The study was undertaken in a period of "normal" natural and economic conditions after the reform. We argue that, assuming there are no major changes in natural factors, changes in the trend of agricultural productivity and production may be ascribed to the impact of policy reforms and institutional changes. To estimate the magnitude of such changes, careful choice of an appropriate model was necessary. Based on the dynamic theory of the optimizing behavior of producers, changes in economic or policy variables can create a new environment that may provide the basis for new optimal decisions. A model that is to be used to evaluate the performance of firms or sectors under such conditions therefore needs to be dynamic as well.

For this reason, changes in the motivation and ability of farmers to use the best techniques and new innovations in the production process were examined using a random coefficient framework. A stochastic Cobb-Douglas production function that incorporates *time-varying inefficiency effects and technological changes* was applied to farm household panel data collected at intervals of years (1994, 1997, 1999 and 2001). As mentioned above, this period was characterized as "normal" in terms of climatic and economic conditions, and it was also a period in which government extension and credit support for transferring technologies and inputs to farmers was at its highest level. So the result should be interpreted to indicate the potential progress (regress) under such normal circumstances. The year 1994 was taken as a base year for comparison of all possible changes in the performance of agriculture in the period analyzed.

The study concentrated on three economically highly important cereal food crops (maize, tef and wheat) that received particular attention in policies aimed at improving food security. To avoid underestimating or overestimating the effects of changes after the reform, three districts representing the country's agricultural potential were carefully selected. Shashamane district represented the zone with high agricultural potential and good market access, Debrebrehan Zuria represented the medium-potential zone and Ankober-Dinqi the low potential zone. Due to constraints of nature, not all crops can grow in all locations. Accordingly, of the three selected cereals only wheat can grow well in Debrebrehan Zuria and only tef and maize in Ankober-Dinqi, whereas all the three cereals grow properly in Shashamane. Detailed analysis and comparisons between the regions were therefore carried out accordingly.

Attempts were made first of all to describe changes in resource endowment (natural and physical capital), factor markets and use of modern inputs in the study period. Overall, while the mean number of livestock units per household was steadily decreasing due to diminishing grazing areas, the mean number of oxen/bull was gradually increasing, but with considerable differences across the ecological zones. It seems that farmers were selectively keeping the more productive classes of livestock. Nevertheless, on average 45% of sample farmers in Ankober-Dinqi, 32% in Shashamane and 13% in Debrebrehan Zuria were without oxen during the study period. Land is generally regarded as scarce given the prevailing farming technology in Ethiopia. There was no major change in the amount of land owned by households over the study period. The mean land size owned per sample household was a little more than 3.0 hectares in Debrebrehan Zuria, and less than 1.5 hectares in the other two locations.

Land and labor transactions in Ethiopia are generally rudimentary. Nevertheless, participation of sample households in the labor market improved slightly, although there was significant divergence among the locations. The highest level of labor transaction activities was observed in Shashamane and the lowest in Ankober-Dinqi. Because of a restrictive land policy, renting and sharecropping were the only alternatives. Interestingly, the proportion of households involved in these transactions was steadily increasing, with the strongest increase being in Shashamane and the smallest in Ankober-Dinqi. Households were more involved in sharecropping transactions than in land renting.

In general, the use of modern inputs was only partially integrated into agricultural production. Recently, there has been a marked increase in the number of farmers using modern inputs (mainly commercial fertilizer and improved seeds) and the average rate of their application. Nevertheless, chemical fertilizer was used only on selected major crops like tef, wheat, maize and barley, and technologies such as improved seeds and better agronomic practices were available only for wheat and maize during the study period. Despite its immense importance in the food sector of the country, the tef production system remained largely traditional. Consequently, the results of the statistical analysis showed that a considerable number of farmers were progressively switching towards the production of wheat and maize (provided that the local ecological conditions are suitable) by reducing the area used for production of tef and other crops (except tef in Ankober-Dinqi). The most intensive use of modern inputs was found in Shashamane and the least intensive in Ankober-Dinqi. This confirms the expectation that farmers in areas with greater agricultural potential and better market access lead the field in terms of adopting improved inputs/modern practices since it is most profitable and involves less risk.

Perhaps the most important findings that emerged from the micro-level analysis were the remarkable growth of output in all the selected production lines and the diverse changes in the TE, TCH and TFP in each case. On average, output of tef was growing at the rate of 14.9%, wheat at 9.6%, and maize at 18% per annum. In the case of wheat alone, however, both TE and TCH were growing positively, at the rate of 3.9% and 4.7% respectively. As a result, an appreciable growth rate in TFP (8.6%) was recorded in this crop, which accounted for the largest share (88.7%) of growth in wheat output. On the other hand, the direction of change in TCH and TE in the other two crops was the opposite. In tef production, TE was increasing significantly at a rate of 3.8%, but its TCH deteriorated (-5.27%) much faster, causing TFP of tef to fall at a rate of 1.45% per annum. Consequently, all the growth recorded in tef output was derived from increased input use, mostly from chemical fertilizer, because the marginal contribution of human and oxen labor was diminishing and the land size used for tef was almost constant in that period. Conversely, while TCH in maize production was progressing at a rate of 10.8% per annum, farmers became gradually more inefficient (TE was declining at 2.9% per annum). However, because the growth in TCH outweighed the decline in TE, TFP for maize was growing at a rate of 7.9% and contributed greatly (44%) to output growth.

It is worth mentioning that, on average, the TE for wheat and tef farmers towards the end of the study period was more than 90%. Thus, additional output gain from improvements in the TE of farmers of these crops given the prevailing technology will be limited. Nevertheless, as the mean TE of maize farmers was about 76%, the gain from improvement of their TE will be immense. Moreover, increasing use of traditional inputs will have limited potential to increase production, as elasticities of labor and oxen use were decreasing (particularly in tef and wheat production); land is limited and is becoming increasingly scarcer.

We should also appreciate the variation in potential of ecological zones that can provide important development policy implications. Close inspection of the variations in the performance of the production lines by agro-ecological potential showed that the TFP of all crops in Shashamane was superior to that in the other districts. In contrast, while the wheat productivity gap between Shashamane and Debrebrehan was decreasing, the productivity gap in maize production between Shashamane and Ankober-Dinqi was increasing in the period analyzed. TFP of tef was declining at a similar rate in both Shashamane and Ankober-Dinqi. The convergence of productivity of wheat in these two districts was due to the increasing use of improved varieties of wheat and fertilizer by the farmers in Debrebrehan Zuria, given that it was initially at a much lower level compared to levels used by farmers in Shashamane, who had already more experience in the use of such inputs. On the other hand, the increasing discrepancy of maize productivity between the locations, regardless of increased use of improved seeds, may suggest the need to specialize in maize production where the potential for improvement is higher.

The results from the analysis of returns to scale (RTS) indicate that the tef and maize production lines in particular were not scale-efficient using the prevailing technology. While tef production experienced decreasing RTS (i.e., over-utilization of resources), maize production experienced increasing RTS (i.e., less than optimum resources used). This suggests the economic importance of reallocating resources to a greater extent towards production of maize, and less towards tef in order to improve scale efficiency and enable farmers to increase food grain output.

As already mentioned, despite the notable progress as regards TCH in maize production, the TE of maize farmers declined by a mean of 15%, which resulted in a loss of about 24% of the potential output at the end of the study period. This clearly demonstrates that market liberalization alone does not always guarantee more efficient production. This result encouraged us to undertake further research to identify the limiting factors using a *Technical Inefficiency Effects* model nested in the stochastic production function. The result generally suggests that TE in maize declined mainly because the changes in terms of institutions and farmers' managerial skills were not in parallel with improvements in maize technology. Whereas household education, access to credit and size of livestock owned positively and significantly influenced the TE of maize farmers, unreliability of the maize price, distance from the main road (remoteness) and the household's dependency ratio negatively affected it.

At the macro level, the study also proceeded to elucidate whether prices of major food products became more unstable after the reform and had a significant impact on modern input use and food grain production. Obviously, market liberalization is repeatedly advocated to boost efficiency and productivity. We also argue, however, that fully liberalizing prices where markets are poorly developed and production is entirely rain-fed could increase the risk of price instability. This in turn would reduce incentives for farmers to use modern inputs and ultimately aggravate food insecurity problems. In this case, both normal and adverse situations in the natural and economic environment were taken into account. Information from the analysis would highlight the prospects for sustainable use of modern inputs, and hence food self-sufficiency and agricultural transformation.

The study therefore tried to estimate the extent of price volatility in major food crops using a *Generalized Autoregressive Conditional Heteroscedasticity* (GARCH) model on monthly time-series data after the reform. The result showed that prices of major food products in Ethiopia tended to swing more severely after the reform than before. Accordingly, both the short-run (monthly) and the long-run (seasonal and annual) volatility measures were statistically highly significant at the 1% level. In particular for wheat and maize, the volatility shocks were persistent. It is worth mentioning that these two crops were those for which improved varieties and chemical fertilizer are predominantly used. The volatility in the aggregated prices of major cereals was also significant at the 10% level.

This result prompted us to go on to estimate the impact of price instability on the use of modern inputs. The analysis was based on annual time-series data collected from different sources. The existence of cointegration and endogeneity between the quantities of modern inputs used and the volume of food grain produced suggested that it would be preferable to use a *Vector Error correction model* (VECM). The results indicated that instability in product prices and unpredictable weather changes were the most strongly determining factors in that they had a significant discouraging effect on the use of modern inputs by small farmers and consequently adversely affected food grain supply. As suggested by the error correction terms, it takes about 3.5 years (28% per annum) to restore the deviation in modern input use to its equilibrium level, while output supply takes about 2 years (50% per annum). This can be considered a long time.

Overwhelmingly, farmers were trapped by twin problems, very low product prices in good harvest years and very low yield in bad years, both effects leading to poor farm returns and hence persistent food insecurity problems. On the one side, intensifying use of modern inputs could not guarantee farmers additional income, as prices of their products fell in successive years (despite normal weather conditions) to the extent that they were unable to pay back they debts they had incurred to purchase the inputs. On the other side, low levels of modern input use (plus bad weather) deprived farmers of adequate output. Then, the poor farmers (as net purchasers of food) were confronted with high food prices, and thus became increasingly vulnerable to hunger.

In conclusion, there is evidence that the economic reforms and institutional changes induced greater use of modern inputs (mainly chemical fertilizer and improved seeds) at least in some food crops in the study period. Nevertheless, its impact on technical efficiency improvement and technological progress was not uniformly positive in the production lines considered. Results from the household-level research confirmed that there is considerable potential for agricultural growth given a normal climatic and economic environment. However, due to a lack of improved inputs for some crops, and the absence of sustainable use for others, the overall impact of the reform was unable to bring about major changes in food production at the national level, implying a negligible aggregate supply response. Among the multifaceted problems, erratic rainfall, unstable product prices, lack of adequate technological innovations and inadequate government support were the major limiting factors. It was observed that a free market does not in itself guarantee any benefit to farmers from using improved and modern production systems.

Overall, although the emphasis on agriculture is consistent with poverty reduction and improving food and nutrition security, market liberalization and technology transfer in the course of Ethiopia's agricultural transformation had only limited success. As a result, it has not been possible to satisfy the food demand of the population and consequently the country has been increasingly dependent on food imports, consisting mainly of aid and some purchased, in an effort to fill the gap created by the shortfalls in food production. This suggests that the present agricultural development policies and strategies for Ethiopia need to be revised to secure the food requirement of the population and enhance overall economic growth.

7.2 Policy Implications

It has been noted that there is considerable potential for producing enough food locally and for promoting agricultural transformation. Investing in agriculture to raise agricultural productivity, especially among small farmers, should therefore be given the highest priority to achieve a secure food supply for Ethiopian society and to foster pro-poor economic growth. This requires increased and sustainable use of modern inputs, which in turn depends on wider adoption of improved technologies, access to markets and prevention of land degradation. Based on the results of this study, more specific proposals are made for policy interventions.

Contrary to our expectations from market liberalization and institutional changes, the fall in the TE of maize and technical regress in tef should attract policy attention. It implies the need to further promote education for the rural households, better access to credit through improving rural financial markets, improvement in rural infrastructure (mainly roads and market centers). Improving transport infrastructure in particular enhances farmers' access to market information, and improves the performance of domestic market and its integration at regional and international levels.

Given that tef is one of the most widely consumed crops, and despite the fact that it uses the largest quantities of inputs including chemical fertilizer, technological deterioration and the fall in TFP of tef severely hamper the food sector of the economy. TE of tef farmers has almost reached its peak under the existing technology; hence, there is little scope for gaining from efficiency improvements. Research to develop improved tef production methods is therefore strongly recommended. In the meantime, as both maize and tef were not scale-efficient given the current technology, reallocation of more resources in favor of maize production rather than tef could improve total farm output. Based on the results for the trends of the elasticities of the inputs, increasing use of human and oxen labor in maize but reducing it in tef would be recommended. Determining the exact amount of the resources to be transferred would require further research and it depends on how technological innovations change over time.

Although progress as regards the TCH and TE of wheat production was impressive, productivity was far below the international standard. Sustainable use of modern inputs should therefore be encouraged, in conjunction with further research to generate better technologies.

Overall, the quickest solution for improving the poor performance of Ethiopian agriculture and alleviating food insecurity is to further encourage the small-scale farmers to increase their use of modern inputs. Chemical fertilizer and improved seeds in particular, coupled with better soil conservation techniques, have enormous potential to boost production. Nevertheless, it was found that severe fluctuations in product prices and unreliable climatic conditions increasingly exposed small farmers to incalculable income risks and sometimes loss of assets. This significantly hindered farmers from using modern inputs. Because poor farmers are risk-averse, they should be protected from significant income loss due to price shocks until they become economically better able to sustain themselves.

Therefore, genuine government intervention to stabilize the food grain system, at least in the short-run until the domestic market is sufficiently developed, should not be considered inconsistent with market liberalization. Small farmers have to be protected from price risks until their management skill is well developed to properly operate in a free market system, and until they economically become better off to withstand external and internal shocks. In a poor market and subsistence economy, it is unrealistic to think of options like future contracts as a strategy to protect farmers against an intra-year fall in revenue. To achieve a given level of price stability and secure food for the nation, public buffer stocks should not be discouraged in the short run.

In the long run, development of market information, storage facilities and increased use of irrigation could curb the problem of seasonal and annual price and output fluctuations. In addition to its price stabilizing effect, improved storage facilities (on and off-farm) can help reduce post harvest losses and increase returns to farmers and traders and significantly contribute to the overall food availability. It is also of paramount importance to support research and improve its effectiveness to develop processing technologies, and to encourage the private sector to engage in the food processing industries to increase local demand for agricultural products and assure its supply chain. This would greatly mitigate instability in the food sector and at the same time promote agro industrialization. Furthermore, developing input markets, especially for fertilizer and improved seeds, is crucial. The government should allow and encourage full participation by private traders so as to improve the supply and distribution of inputs. Given the low level of cash income that farmers have at their disposal, promoting micro-finance institutions accessible to small farmers could play an immense role in the use of modern inputs and in agricultural transformation as a whole.

Zusammenfassung und Politikfolgen

Zusammenfassung

Ungeachtet vieler Kontroversen über die Rolle des Agrarsektors im Entwicklungsprozess eines Landes gibt es viele Beispiele dafür, dass die Landwirtschaft eine wichtige Rolle beim wirtschaftlichen Wachstum und strukturellen Wandel heutiger Industrie- und Schwellenländer gespielt hat. Ein Vergleich zwischen den asiatischen Tigerstaaten und Ländern in Sub-Sahara Afrika z.B. zeigt, dass die Zunahme landwirtschaftlicher Einkommen die wichtigste Erklärungsgröße darstellt, mit der die sehr unterschiedlichen Entwicklungspfade erklärt werden können. Eine mangelnde Beachtung oder Vernachlässigung des Agrarsektors beeinträchtigt die Nahrungsmittelproduktion, unterminiert die Ernährungssicherheit und behindert die Freisetzung von Ressourcen, die zur Expansion der nicht-landwirtschaftlichen Sektoren im Prozess struktureller Transformation benötigt werden.

In Äthiopien gab es in den vergangenen fünfzig Jahren drei politische Regime, die sich in Bezug auf ihre entwicklungspolitischen Strategien deutlich unterschieden. Das feudal-bürgerliche Regime verfolgte eine Mischung aus „inward-oriented" und „outward-oriented" Politik, während das sozialistische eine streng nach innen gerichtete Politik wählte. Beiden war jedoch gemeinsam, dass sie sich auf die Entwicklung des Industriesektors konzentrierten und den landwirtschaftlichen Sektor vernachlässigten. Zusammen mit periodisch auftretenden Trockenperioden und dem Bürgerkrieg führte diese einseitige Präferenz in beiden Fällen zum Schaden der gesamten Volkswirtschaft, mit sehr gravierenden Auswirkungen für den Agrarsektor. Diese verheerende Entwicklung führte schließlich dazu, dass die gegenwärtige Regierung die Notwendigkeit wirtschaftlicher Reformen und institutioneller Änderungen erkannte und einen Transformationsprozess der vormals planwirtschaftlich gelenkten Volkswirtschaft hin zu einem marktwirtschaftliches System einleitete. Kernpunkte dieses Prozesses waren die Liberalisierung von Märkten, Privatisierungen von Staatsbetrieben und der Aufbau eines Finanzsektors. Im Bereich der Landwirtschaft setzte eine Förderung von Forschung und Beratung zur Stärkung der ruralen Wirtschaft ein.

Für eine nachhaltige Armutsreduzierung ist das Wachstum des Agrarsektors von zentraler Bedeutung. Deshalb hat die Regierung der Landwirtschaft höchste Priorität und eine führende Rolle im Entwicklungsprozess eingeräumt und mit der Strategie „Agricultural Development-Led-Industrialisation" (ADLI) ein Programm konzipiert, das auf ländliche Entwicklung und exportorientiertes Wachstum setzt. Im Zentrum dieser Strategie steht die Erhöhung der landwirtschaftlichen Produktivität von Kleinbauern, die die Subsistenzproduktion sicherstellen, Überschüsse zur Versorgung der nicht-landwirtschaftlichen Bevölkerung mit Nahrungsmitteln produzieren und den Bedarf der Industrie an Rohstoffen befriedigen soll. Dieser radikale Politikwechsel begleitet von institutionellen Reformen soll zu einer nachhaltigen wirtschaftlichen Entwicklung und einer Verbesserung des Lebensstandards der gesamten Gesellschaft führen. Die Beantwortung der Frage, inwieweit dieses Ziel erreicht wurde und welche Konsequenzen sich aus der bisherigen Entwicklung ergeben, wird im Folgenden behandelt.

Das Potenzial des Agrarsektors als Wachstumsmotor ist eng verknüpft mit dem Funktionieren von Produkt- und Faktormärkten, was eine Grundvoraussetzung für die effiziente Allokation von Ressourcen ist. Zwischen den zu erwartenden und tatsächlichen Wachstumsraten besteht aufgrund institutioneller Schwächen und Marktversagen in Äthiopien jedoch eine große Diskrepanz. Die Initialisierung von Wachstumsprozessen durch Produktivitätssteigerungen ist eine der wichtigsten und zugleich schwierigsten Aufgaben für viele Entwicklungsländer. Die vorliegende Arbeit versucht deshalb, die Effekte politischer und institutioneller Reform auf die landwirtschaftliche Entwicklung zu bewerten, um daraus Wirkungsmechanismen zu erkennen und Politikempfehlungen zu geben.

Die Arbeit vergleicht zunächst die makroökonomische Situation Äthiopiens vor und nach der Reform und stellt die Bedeutung des Agrarsektors in der Volkswirtschaft dar. Es wird deutlich, dass die Staatsausgaben in den Bereichen Landwirtschaft, Erziehung und Straßenbau mit der Reform deutlich gestiegen sind, während die Ausgaben für den Industriesektor deutlich rückläufig waren. Dabei wurde seitens der Regierung antizipiert, dass der private Sektor einen großen Beitrag zum Ausbau der Industrie leisten würde, was aber aufgrund eines langsamen Privatisierungsprozesses und Rechtsunsicherheit v.a. im Bereich der Eigentumsrechte nicht im nötigen Umfang geschah.

Um mehr Licht auf diese Entwicklung zu werfen, wurde ein einfaches *log lineares* Wachstumsmodell angewendet, um Trends in den verschiedenen Sektoren zu schätzen. Obwohl leicht positive Wachstumstrends in allen Sektoren beobachtet werden konnten, stagnierte das BIP und die Nahrungsmittelproduktion pro Kopf. Gleichzeitig stiegen die Auslandsschulden schneller als je zuvor. Der Anteil des Agrar- und Industriesektors blieb nahezu unverändert, während der Dienstleistungssektor an Bedeutung gewann und den größten Beitrag zum Wachstum (>60%) leistete. Stagnation im Industriesektor ist vor allem auf die negativen Wachstumsraten des Kleingewerbes zurückzuführen, während Großbetriebe und mittelständische Unternehmen stark wuchsen. Der Gesamtanteil des Industriesektors betrug 10%, aber sein relativer Beitrag zum Wirtschaftswachstum fiel um die Hälfte von 25% auf 12% nach der Reform. Der Anteil der Landwirtschaft am BIP sank um 10% nach der Reform (im Einklang mit dem entwicklungstheoretischen Stufenmodell). Dabei blieb die Rolle des Agrarsektors für das gesamtwirtschaftliche Wachstum hinter den Erwartungen zurück. Es trug lediglich 17% zum BIP Wachstum bei im Vergleich zu 9% vor der Reform. Der Anteil außerlandwirtschaftlicher Beschäftigung nahm lediglich von 10% auf 15% der Gesamtbeschäftigung in den letzten 15 Jahren zu und dieser Trend verlangsamte sich sogar nach der Reform, was hauptsächlich- im Gegensatz zu ADLI's Intention- auf die Entwicklung hin zu kapitalintensiver Produktion zurückzuführen ist.

Obwohl das Exportvolumen zunahm, kam es zu einer Verschlechterung der Handelsbilanz. Dies ist auf eine weitaus größere Zunahme der Importe (v.a. Konsumgüter) im Verhältnis zu den Exporten und auf eine sinkende Tendenz der Weltmarktpreise für Äthiopiens Hauptexportprodukte zurückzuführen. Ein weiterer außenwirtschaftlicher Trend ist die zunehmende Abhängigkeit des Agrarsektors und der verarbeitenden Industrie von importierten Rohstoffen. Diese Entwicklungen sind als sehr problematisch einzuschätzen, da die Zahlungsbilanz nur über steigende Auslandverschuldungen ausgeglichen werden kann. Sie erscheinen daher aus der Perspektive der Entwicklungstheorie als nicht nachhaltig und wurden im Rahmen dieser Arbeit sowohl auf der Mikro- als auch auf der Makroebene näher untersucht.

Auf der Haushaltsebene (Mikroebene) wurde der Frage nachgegangen, inwieweit sich die technische Effizienz (TE), die Technologie (TCH), die totale Faktorproduktivität (TFP) und die Intensität des Inputaufwandes ändern und welchen Beitrag zum Ertragswachstum sie leisten. Die empirische Datenerhebung fand in Perioden „normaler“ klimatischer und ökonomischer Bedingungen statt. Bei der Analyse wurde unterstellt, dass unter der Annahme der Abwesenheit bedeutender natürlicher Veränderungen eine Kausalität zwischen Änderungen in Produktions- und Produktivitätstrends in der Landwirtschaft und den Politikreformen besteht. Um diese Einflüsse zu quantifizieren ist die sorgfältige Auswahl eines Modells entscheidend. Der dynamischen Theorie vom gewinnoptimierenden Produzentenverhalten folgend, könnten Veränderungen in exogenen Variablen zu neuen Voraussetzungen führen, die wiederum zu einem neuen optimalen Verhaltensgleichgewicht führen. Ein dynamisches Modell, was die Leistung von Firmen und Sektoren unter solchen Bedingungen erfassen kann, ist somit wichtig. Deshalb wurden Veränderungen in der Motivation und die Möglichkeit von Bauern, die beste Technik zu nutzen und neueste Technik anzuwenden anhand des random coefficient Konzept getestet. Dabei kam eine stochastische Cobb-Douglas Produktionsfunktion, die *Ineffizienzeffekte zu verschiedenen Zeitpunkten und technologischen Wandel* inkooperiert, zur Anwendung.

Datengrundlage bilden Haushalt- Paneldaten die in den Jahren 1994, 1997, 1999 und 2001 erhoben wurden. In diesem Zeitraum waren die Aufwendungen für staatliche Beratung und Kreditprogramme zum Transfer von Technologien und Inputs am höchsten. Die Ergebnisse können somit als Indikator für Fortschritte (Rückschritte) in der Entwicklung interpretiert werden. Das Jahr 1994 wurde als Referenzjahr für Vergleiche mit allen möglichen Änderungen in der landwirtschaftlichen Entwicklung gewählt.

Die Untersuchung konzentriert sich auf die drei ökonomisch wichtigsten Getreidearten in Äthiopien, namentlich Mais, Tef und Weizen. Um eine

Überschätzung oder Unterschätzung der Effekte von Veränderungen nach der Reform zu vermeiden, wurden drei verschiedene Regionen ausgewählt, die das landwirtschaftliche Potential Äthiopiens repräsentieren. Shashemene Distrikt repräsentiert die Gegenden mit hohem Potential und gutem Marktzugang, Debrebrehan Zuria Gegenden mit mittlerem Potential und Ankober-Dinqi mit schwachem Potential. Die geoklimatischen Bedingungen in diesen Distrikten sind sehr unterschiedlich und es werden nicht alle untersuchten Getreidesorten in allen Regionen gleichermaßen angebaut. Nur Weizen in Debrebrehan Zuria und nur Tef und Mais in Ankober-Dinqi haben gute Wachstumsbedingungen, während alle drei Arten gut in Shashamane wachsen.

Verschiedene Ansätze wurden verfolgt, um Änderungen in der Ressourcenausstattung (natürliches und physisches Kapital), Faktorausstattung und den modernen Materialeinsatz im Untersuchungszeitraum zu darzustellen. Während die mittlere Viehbestandsdichte pro Haushalt wegen abnehmenden Weideflächen ständig rückläufig war, stieg die durchschnittliche Anzahl der gehaltenen Zugtiere (Ochsen, Bullen) ständig an, allerdings mit erheblichen Unterschieden zwischen den Regionen. 45% der getesteten Betriebe in Ankober-Dinqi, 32% in Shashemene und 13% in Debrebrehan Zuria hielten keine Ochsen. Landressourcen werden allgemein als knapp eingestuft unter gegebener Technologie. Im Betrachtungszeitraum gab es kaum Änderungen in der durchschnittlichen Eigentumsfläche pro Haushalt, so entfielen 3 ha in Debrebrehan Zuria und weniger als 1,5 ha in den beiden anderen Regionen auf einen Haushalt.

Land- und Arbeitsmärkte sind kaum entwickelt in Äthiopien. Trotzdem hat sich die Teilnahme der getesteten Haushalte am Arbeitsmarkt etwas verbessert, aber wiederum mit großen Unterschieden zwischen den Regionen. Die höchste Arbeitsmarktaktivität wurde in Shashemene beobachte, die niedrigste in Ankober-Dinqi. Auf dem Landmarkt dominiert wegen einer restriktiven Landpolitik die Pacht als weitgehend einzige Alternative.

Über die Anwendung moderner Inputs lässt sich verallgemeinernd sagen, dass diese nur teilweise ins landwirtschaftliche Produktionssystem integriert sind. In jüngerer Zeit stieg jedoch die Zahl derjenigen Bauern, die v.a. von Handelsdünger und verbessertem Saatgut Gebrauch machten. Mineraldünger wurde hauptsächlich für Tef, Weizen, Mais und Gerste verwendet. Verbessertes Saatgut und fortschrittliche pflanzenbauliche Praxis kam nur für Weizen und Mais zur Anwendung. Trotz seiner immensen Bedeutung für die Ernährung des Landes, blieb die Tefproduktion zum großen Teil von traditionellen Anbaumethoden geprägt. Die statistische Analyse zeigt, dass eine beträchtliche Anzahl von Bauern mit der Ausnahme von Ankober-Dinqi ihre Produktion auf Weizen und Mais umgestellt haben. Die intensivsten Bewirtschaftungsformen waren dabei in Shashamane und die extensivsten in Ankober-Dinqi zu finden. Das bestätigt die These, dass Bauern in potentialreicheren Gegenden und mit besserem Marktzugang führend in der Adaption moderner Anbausysteme sind.

Vielleicht das bedeutendste Ergebnis der Mikroanalyse waren erhebliche Produktionssteigerungen in allen untersuchten Produktionslinien, aber jeweils sehr unterschiedlichen Änderungen in TE, TCH, und TFP. Im Durchschnitt verzeichnete Tef jährliche Ertragszuwächse von 14,9%, Weizen von 9,6%, und Mais von 18%. TE and TCH war aber nur bei Weizen mit 3,9% bzw. 4,7% steigend. Im Gegensatz dazu stieg zwar die TE in der Tefproduktion um 3,8%, aber TCH fiel mit 5,27% schneller, was zu einer Minderung der TFP um 1,45% pro Jahr führte. Folglich sind die Zuwächse in der Tefproduktion ausschließlich auf erhöhten Materialeinsatz (v.a. Handelsdünger) zurückzuführen, da der Grenzertrag der menschlichen und tierischen Arbeitskraft beständig abnahm und die genutzte Landfläche konstant blieb. Umgekehrt, während die TCH in der Maisproduktion um 10.8% zunahm, sank die Effizienz der Bauern um 2.9% pro Jahr. Da der Anstieg in TCH das Sinken der TE ausglich, verzeichnete die TFP für Mais ein Wachstum von 7.9% und trug somit stark zum Ertragswachstum bei (44%). Die TE der Weizen- und Tefbauern am Ende des Untersuchungszeitraumes betrug im Durchschnitt 90%. Zusätzliche Ertragszuwächse und zusätzliches Ertragswachstum sind deshalb limitiert. Da aber die anfängliche TE der Bauern nur bei 76% lag, waren die Zuwachsgewinne groß. Es kann konstatiert werden, dass ein steigender Einsatz traditioneller Inputs nur begrenzte Wachstumspotentiale bietet, da die Elastizitäten für Arbeit und tierische Zugkraft abnehmende Tendenz aufweisen. Landressourcen sind begrenzt und werden zunehmend knapper.

Es ist wichtig bei der Formulierung von entwicklungspolitischen Strategien, die unterschiedlichen Potentiale der ökologischen Zonen zu beachten. Eine genaue Untersuchung der verschiedenen produktionsrelevanten natürlichen Potentiale zeigte, dass die TFP aller Getreidesorten in Shashamane den anderen beiden Standorten überlegen waren. Auf der anderen Seite stieg die Produktivitätslücke zwischen Shashamane und Debrebrehan im Bereich des Weizenanbaus. TFP in der Tefproduktion wies eine gleichermaßen sinkende Tendenz sowohl in Shashamane als auch in Ankober-Dinqi auf. Die Produktivitätskonvergenz für Weizen in diesen beiden Distrikten ist auf den verstärkten Einsatz von verbesserten Sorten und verstärkten Düngereinsatz in Debrebrehan Zuria zurückzuführen. Die zunehmenden Unterschiede in der Produktivität bei Mais zwischen den Gebieten, unabhängig von verstärktem Einsatz verbesserten Saatgutes, deuten auf die Notwendigkeit einer Spezialisierung auf Mais hin, wo Verbesserungspotentiale größer sind.

Die Ergebnisse der Skalenertragsanalyse (RTS) indizieren, dass v.a. die Tef- und Maisproduktion nicht skaleneffizient unter der vorherrschenden Technologie waren. Während die Tefproduktion abnehmende RTS (ein Anzeichen der Bodenübernutzung) verzeichnete, zeigte die Maisproduktion steigende RTS (d.h. suboptimale Ressourcennutzung). Diese Ergebnisse unterstreichen die ökonomische Bedeutung einer Reallokation von Ressourcen in Richtung der Maisproduktion und weniger in Richtung Tefproduktion, um die Skaleneffizienz zu steigern und die Bauern zu höheren Erträgen zu befähigen.

Es wurde bereits erwähnt, dass trotz der bemerkenswerten Fortschritte in TCH bei Mais, die TE im Durchschnitt um 15% sank, was zu einem Verlust von ungefähr 24% des potentiellen Ertrages am Ende der Untersuchungsperiode führte. Das demonstriert eindeutig, dass Marktliberalisierung alleine nicht immer zu effizienterer Produktion führt. Dieses Zwischenergebnis wurde zum Anlass genommen, die limitierenden Faktoren mit Hilfe eines Modells *technischer Ineffizienzeffekte* zu analysieren verschachtelt mit einer stochastischen Produktionsfunktion. Das Resultat deutet darauf hin, dass die TE bei Mais hauptsächlich deshalb abnahm, weil institutionelle Änderungen und das Fachwissen der Bauern nicht in gleichem Maße zunahmen wie die Verbesserung der Maistechnologie. Während Erziehung, Zugang zu Krediten und die Viehbestandsgröße positiv mit der TE korreliert waren, hatten die Entfernung zur nächsten Strasse, schwankende Maispreise und die Nicht-Erwerbstätigen Proportion der Haushalte negativen Einfluss.

Auf der makroökonomischen Ebene wurde der Frage nachgegangen, ob die Volatilität der Preise nach der Reform zunahm und einen Einfluss auf den Gebrauch von modernen Inputs und die Getreideproduktionsvolumina hatte. Marktliberalisierung wird immer wieder in kausalen Zusammenhang mit Effizienz- und Produktivitätssteigerung gebracht. Wir argumentieren, dass eine volle Preisliberalisierung in einem Umfeld mit wenig entwickelten Märkten und komplett regenabhängiger Landwirtschaft das Risiko von Preisinstabilitäten erhöhen kann. Dies wiederum kann zu einem negativen Anreiz führen, moderne Inputs zu verwenden und somit eventuell die Gefahr von Ernährungsunsicherheiten zu verstärken. Um tiefere Einblicke in diese Zusammenhänge zu bekommen, wurden verschiedene Szenarien untersucht.
Dazu wurde das Ausmaß der Preisvolatilität der Hauptfrüchte mittels eines *Generalized Autoregressive Conditional Heteroscedasticy* (GARCH) Modells anhand von monatlichen Zeitreihendaten nach der Reform geschätzt. Die Ergebnisse zeigen, dass die Preise nach der Reform stärker schwankten als vor der Reform. Sowohl die kurzfristigen (monatlichen), als auch die mittel- und langfristigen (jahreszeitlichen und jährlichen) Volatilitätsmesszahlen waren statistisch hoch signifikant im Bereich von 1%. Besonders für Weizen und Mais waren die Volatilitätsschocks persistent. Dies sind diejenigen Früchte, für die auch vorwiegend verbesserte Sorten angebaut und Handelsdünger eingesetzt wurden. Auch für das Aggregat der Getreidesorten konnte eine signifikante Volatilität im 10% Bereich festgestellt werden.

Diese Ergebnisse veranlassten uns dazu, die Auswirkungen dieser Preisschwankungen auf den Einsatz moderner Inputs zu testen. Die Existenz von Kointegration und Endogenität zwischen Materialeinsatz und der Menge produzierter Nahrungsmittel favorisiert den Einsatz eines *Vector of Correction models* (VECM). Die Analyse basiert wiederum auf jährlichen Zeitreihendaten.

Die Ergebnisse zeigen, dass instabile Produktpreise und unvorhergesehene Wetteränderungen die wichtigsten Determinanten waren, die eine signifikant demotivierende Wirkung in Bezug auf den Einsatz moderner Inputs bei den Bauern hatte. Die Fehlerkorrekturgröße lässt darauf schließen, dass die Abweichung moderner Inputs etwa 3.5 Jahre braucht und das Ertragsniveau 2 Jahre, um zu ihrem Ausgangsniveau zurückzukehren. Dies sind lange Zeiträume.

Die Situation in Jahren guter Ernten niedrige Preise und in schlechten Erntejahren gute Produktpreise zu erzielen, führt in beiden Fällen zu schlechten Erlösen und damit anhaltender Nahrungsmittelunsicherheit. Die meisten Bauern befinden sich in einem Dilemma. Einerseits konnten durch eine Intensivierung der Produktion keine anhaltenden zusätzlichen Erlöse garantiert werden, da die Preise in nachfolgenden Jahren sanken, was zu der Situation führte, dass Schulden, die zum Kauf von Saatgut und Handelsdünger aufgenommen wurden, nicht zurückgezahlt werden konnten. Andererseits konnten mit extensivem Einsatz moderner Inputs kein ausreichender Ertrag und somit auch Erlös erzielt werden. Bei hohen Lebensmittelpreisen trat für arme Bauern als Nettokonsumenten zusätzlich eine ernstzunehmende Hungergefahr auf. Dies erklärt die zunehmende Abhängigkeit des Landes von ausländischer Nahrungsmittelhilfe. Außerdem ermutigte die Regierung private Händler, Lebensmittel aus dem Ausland zu importieren um Engpässe zu überbrücken.

Zusammenfassend lässt sich sagen, dass die wirtschaftlichen Reformen und institutionellen Änderungen zu vermehrtem Einsatz von modernen Inputs zumindest bei einigen Nahrungsmitteln geführt hat. Der Einfluss auf die Verbesserung der technischen Effizienz und technologischem Fortschritt sind nicht durchweg positiv. Die Ergebnisse der vorliegenden Arbeit deuten jedoch darauf hin, dass ein erhebliches Potential für wirtschaftliches Wachstum im Agrarsektor besteht. Zu den vielfältigen Problemen zählen vor allem unsteter Regenfall, hohe Produktpreisunsicherheit, ein Mangel an angepassten technologischen Innovationen und unangemessene Regierungsinterventionen.

Obwohl also die Schwerpunktsetzung auf den landwirtschaftlichen Sektor als konsistent mit den Zielen der Armutsreduzierung und der Erreichung von Ernährungssicherheit betrachtet werden kann, hatten Marktliberalisierung und Technologietransfers nur begrenzten Erfolg. Im Gegenteil, die heimische Lebensmittelnachfrage konnte nicht befriedigt werden und das Land ist in zunehmendem Masse abhängig von Importen. Damit kann festgestellt werden, dass die gegenwärtige landwirtschaftliche Entwicklungspolitik Äthiopiens einer Revision bedarf, um die Ernährungssicherheit der Bevölkerung zu sichern und allgemeinwirtschaftliches Wachstum zu befördern.

Folgen für die Politik

Es besteht ein beachtliches Potential für eine Steigerung der nationalen Nahrungsmittelproduktion und einen weitergehenden Strukturwandel in der Landwirtschaft. Deshalb sollten Investitionen zur Produktivitätssteigerung im Agrarsektor besonders unter Kleinbauern hohe Priorität eingeräumt werden. Dazu ist ein verstärkter und nachhaltiger Einsatz moderner Inputs nötig, der wiederum von der Adaption verbesserter Technologien, Marktzugang und der Vermeidung von Landdegradation abhängt. Des Weiteren ist es von höchster Wichtigkeit, den Privatsektor zu fördern, um den Bereich der lebensmittelverarbeitenden Industrie auszubauen. Damit würde die lokale Nachfrage nach landwirtschaftlichen Rohstoffen erhöht, die lokale Wertschöpfung vergrößert und die Versorgungskette gestärkt.

Dem Rückgang der technischen Effizienz in der Maisproduktion und dem technische Rückgang in der Tefproduktion sollten hohe politische Aufmerksamkeit geschenkt werden. Um die technische Effizienz bei Maisbauern zu steigern, sollte die Ausbildung ländlicher Haushalte weiter gefördert, ein besserer Zugang zu Krediten durch den Ausbau ländlicher Finanzmärkte gewährleistet und die Infrastruktur, vor allem Strassen, ausgebaut werden.

In Anbetracht der Tatsache, dass Tef eines der am meisten konsumierten Nahrungsmittel in Äthiopien ist und dass es absolut gesehen die größte Menge an Handelsdünger für sich beansprucht, bedroht dessen technischer Rückgang und der Rückgang seiner technischen Effizienz ernsthaft den gesamten Nahrungsmittelsektor. Da die technische Effizienz unter bestehender Technologie fast seine Spitze erreicht hat, gibt es wenig Raum für Verbesserungen der Effizienz. Deshalb ist eine verstärkte Forschung für verbesserte Produktionsmethoden essentiell. In der Zwischenzeit könnte eine Reallokation von Ressourcen von den skalenineffizienten Früchten Weizen und Tef hinzu Mais erfolgen, um den gesamtlandwirtschaftlichen Ertrag zu erhöhen. Basierend auf den Ergebnissen zu den Elastizitätstrends der Faktoren, wäre eine Expansion menschlicher und tierischer Arbeitskraft in der Maisproduktion und eine Reduktion dieser Faktoren bei Tef empfehlenswert.

Obwohl Fortschritte in der Technologie und technischen Effizienz bei Weizen erreicht werden konnten, bleib die Produktivität weit hinter internationalen Standards zurück. Deshalb sollte weitere Forschung zur Generierung besserer Technologien gefördert werden.

Eine schnelle Lösung, um die schlechte Leistung der äthiopischen Landwirtschaft zu verbessern und Ernährungsunsicherheiten zu umgehen ist die weitere Förderung der Kleinbauern zur Adaption neuer Technologien. Vor allem die Anwendung von Handeldünger und verbessertem Saatgut in Kombination mit verbesserter bodenschonender Bearbeitung hat ein enormes Potential die Produktion zu steigern. Es zeigte sich, dass die Bauern durch starke

Produktpreisfluktuationen und unstete Wetterverhältnisse einem hohen Risiko von Einkommens- und teilweise auch Vermögensverlusten ausgesetzt waren, was wiederum den Einsatz moderner Inputs demotivierte. Da arme Bauern ein stark risiko-averses Verhalten aufweisen, sollten sie in einer Übergangszeit vor allzu großer Preisvolatilität geschützt werden. Ein staatliches Eingreifen zur Preisstabilisierung auf dem Getreidemarkt, bis dieser Markt in ausreichendem Masse entwickelt ist, sollte nicht als inkonsistent mit freiem Handel betrachtet werden. Preisausgleichsmechanismen für volatile Märkte wie etwa Warenterminbörsen, die den Bauern mehr Planungssicherheit geben könnten, sind unter gegebenen Umständen keine Alternative. Um eine gewisse Stabilität zu erreichen, sollte auf kurze Sicht deshalb die staatliche Lagerhaltung weiter gefördert werden. Auf lange Sicht muss ein für alle zugängliches Marktinformationsystem und ein Netz von Getreidespeichern geschaffen werden. Weiterhin könnte die Verbreitung von Bewässerungslandwirtschaft dazu beitragen, saisonale und jährliche Preisschwankungen auszugleichen. Zusätzlich ist es wichtig, funktionierende Märkte für Vorleistungsgüter wie v.a. Düngemittel und Saatgut zu schaffen. Die Regierung sollte dabei privaten Händlern mehr unternehmerischen Freiraum gewähren und diese fördern. Schließlich kann der Aufbau von Mikrofinanzinstitutionen für Kleinbauern einen erheblichen Beitrag zur Überwindung von Liquiditätsproblemen und Investitionsengpässen leisten und somit dem verstärkten Einsatz moderner Inputs, dem Strukturwandel in der Landwirtschaft und somit einer positiven gesamtwirtschaftlichen Entwicklung Äthiopiens dienen.

References

Abdulai, A. and W. Huffman. 2000. Structural Adjustment and Economic Efficiency of Rice Farmers in Northern Ghana. *Journal of Economic Development and Cultural Change,* Vol. 48, No. 3, 503-519.

Abebe, H.G. and Mulat, D. 2003. Endowment Profiles and Adoption of Agricultural Technologies: Distributional Dimensions and Impacts on Direct Production Entitlement, In *Technological Progress in Ethiopian Agriculture, Proceedings of the National Workshop on Technological Progress in Ethiopian Agriculture,* Mulat, D., Alemu, M. and Assefa, A. (eds), Addis Ababa, Ethiopia.

Abebe, S. 2004. Addressing the Problem of Poverty in Ethiopia and Policy Options, *htt://eea.ethiopiaonline.net/Econ-Foc/ef2-5/abeshim.htm.* Accessed in March 2004.

ADB. 2001. Asian Development Bank Report on Asian Environment Outlook, *http://www.adb.org.* Accessed April 2003.

Adelman, I. and S. Vogel. 1992. The Relevance of ADLI for Sub-Saharan Africa, In *Industrialization based on Agricultural development: African Development Perspectives Yearbook 1990/91:* Research Group on African Development Perspectives, Bass et al (eds.), Bremen.

Ahmed, R. 1989. Pricing Principles and Public Intervention in Domestic Markets: n *Agricultural Price Policy for Developing Countries*, Mellor, J., and R. Ahmed (eds.), Johns Hopkins University Press.

Aigner, D., C. A. Loevell and P. Schmidt. 1977. Formulation and Estimation of Stochastic Frontier Production Function Models. *Journal of Econometrics,* 6: 21-37.

Alemayehu, G. 1999. Determinants of Aggregate Primary Commodity Export Supply from Africa: An Economic Study. *Ethiopian Journal of Economics*, Vol. 1, EEA.

Alemayehu, G. 2001. Macroeconomic Policy and Agriculture in Ethiopia, unpublished.

Amare, G. 1978. Zonation of Highlands of Tropical Africa: The Ethiopian Highlanlds, ILCA, Addis Ababa.

Anderson, K. 1987. On Why Agriculture Declines with Economic Growth. *Agricultural economics*, Amsterdam: 195-207

Arndt, H.W. 1981. Economic Development: A Semantic History. *Journal of Economic Development and Cultural Change,* 29(74), 457-66.

Assefa, A. 1995. Analysis of Production Efficiency and the Use of Modern Technology in Crop Production: A Study of Smallholders in the Central Highlands of Ethiopia, Ph.D. diss., University of Hohenheim, Kiel: Wissenshaftsveralg Vauk.

Baillie, R. and T. Bolleerslev. 1989. The Message in Daily Exchange Rates: A Conditional Variance Tale. *Journal of Business and Economic Statistics* 7: 297-305.

Baillie, R.T., T. Bollerslev and O. Mikkelsen. 1996. Fractional Integrated Generalized Autoregressive Conditional Heteroscedacity. *Journal of Econometrics* 73: 5-59.

Bale, M. 2001. Reaching the Rural Poor in a Globalizing. *World Bank Paper for Expert Group Meeting on Globalization and Poverty Reduction: Can the Rural Poor Benefit from Globalization,* World Bank, Washington D.C.

Barkoulas, J., C. Walter and O. Joseph. 1997. Fractional Dynamics in International Commodity Prices. *The Journal of Future Markets* 17-2: 161-189.

Barrett, C. and C. Michael. 1994. Does It Take More than Market Liberalization? The Economics of Sustainable Agrarian Growth and Transformation. *Working Paper Series on Development at a Crossroads: Uncertain Paths to Sustainability*, No. 4, Global Studies Research Program, University of Wisconsin, Madison.

Battese G.E .1992. Frontier Production Functions and Technical Efficiency: A Survey of Empirical Applications in Agricultural Economics. *Journal of Agricultural Economics,* 7: 185-208.

Battese, G.E and T. G. Coelli. 1995. A Model of Technical Efficiency Effects in a Stochastic Frontier Production Function for Panel Data. *Empirical Economics* 20: 325-332.

Battese, G.E and T.J. Coelli. 1992. Frontier Production Functions, Technical Efficiency and Panel Data: with Application to Paddy Farmers in India. *Journal of Production Analysis,* 3: 153-169.

Battese, G.E, and S. Broca. 1997. Functional Form of Stochastic Frontier Production and Models for Technical Inefficiency Effects: A Comparative Study for Wheat Farmers in Pakistan. *Journal of productivity Analysis* 8: 395-414.

Belay, E. 1997. The New Extension Strategy: In *Economic Focus,* Alemayehu, G. (ed.), Ethiopian Economic Association, Vol 1, No. 2.

Berhanu, N. and Seid, N. 1999. Performance of the Ethiopian Economy 1991-1998. In *The Ethiopian Economic Performance and Evaluation: Proceedings of the Eighth Annual Conference of the Ethiopian Economy*, Alemayehu G. and Berhanu N. (eds.), Addis Ababa, Ethiopian Economic Association.

Berry, D. and L. Bernard .1996. *Statistics: Theory and Methods*, 2nd edition, USA, ITP.

Beyene, T. 2002. Differential Adoption of Agricultural Technologies and its Implications for Policy Choice between Equity and Growth, In *International Conference on Contemporary Development Issues in Ethiopia,* Western Michigan University, USA.

Beynon, J and J. Stephen. 1992. Review Paper-Market Reform and Private Trade in Eastern and Southern Africa, *Food Studies Group Working Paper* 6. University of Oxford, Queen Elizabeth House.

Beynon, J., J. Stephen and Y. Shujie. 1992. Market Reform and Private Trade in Eastern and Southern Africa. *Food Policy* 17: 399-408.

Beynon. J. 1988. Pricism versus Structuralism in Sub-Saharan African Agriculture. *An Essay Submitted to the Agricultural Economics Society*, Prize Essay of 1988, American Economic Association

Binswanger, H. 1989. The Policy Response of Agriculture, In *Proceedings of the World Bank Annual Conference on Development Economics*, Staley Fischer and Dennis de Tray (eds), World Bank.

Birdsall, N., R. David and S. Richard. 1995. Inequality and Growth Reconsidered: Lessons from East Asia. *World Bank Economic Review* 9(3): 477-508.

Birnberg, T. B. and S.A. Resnick. 1975. *Colonial Development*. New haven, Yale University Press.

Block, S. and P. Timmer. 1994. Agriculture and Economic Growth: Conceptual Issues and the Kenyan Experiences Unpublished.

Boger, S. 2001. Agricultural Markets in Transition- An Empirical Study of Contracts and Transaction Costs in the Polish Hog Sector. *Institutional Change in Agriculture and Natural Resources,* (4). Aachen : Shaker Verlag.

Bollerslev, T. and E. Ghysels .1996. Periodic Auto Regression Conditional Heteroscedacity. *Journal of Business and Economic Statistics,* 14: 139-151.

Bollerslev, T. and O. Mikkelsen. 1996. Modeling and Pricing Long Memory in Stock Market Volatility. *Journal of Econometrics* 73: 151-184.

Bollerslev, T. 1986. Generalized Autoregresive Condtional Heteroscedacity. *Journal of Econometrics,* 52: 307-327.

Bond, M. 1983. Agricultural Responses to Prices in Sub-Saharan African Countries. *International Monetary Fund Staff Paper, 30,* 703-26.

Bravo-Ureta, E. Boris and E. A. Pinheiro. 1997. Technical, Economic, and Allocative Efficiency in Peasant Farming: Evidence from the Dominican Republic. *The Development Economics* XXXV-1: 48-67.

Breidt, F., C. Nuno, de L. Padero. 1998. The Detection and Determination of Long Memory in Stochastic Volatility. *Journal of Econometrics* 83: 325-348.

Brüntrup, M. 1997. Agricultural Price Policy and its Impact on Production, Income, Employment and the Adoption of Innovations: A Farming System Based Analysis of Cotton Policy in Northern Benin, Ph. D. dess. University of Hohenheim. *Development Economics and Policy,* Heidhues F. (ed), Frankfurt, Europäischer Verlag der Wissenschaften.

Buchenrieder, G. 1996. The Role of Rural Finance for Food Security of the Poor in Cameroon. *Development economics and policy,* Ph. D. diss., University of Hohenheim, Frankfurt, Peter Lang.

Chenery, H. B. 1960. Patterns of Industrial Growth. *American Economic Review* 50, 624-54 (232)

Chiang, A. 1984. *Fundamental Methods of Mathematical Economics*, 3rd editions, Singapore, McGRAW-HiLL press.

Chuta, E. and L. Carl. 1990. Rural Small Scale Industry: Empirical Evidence and Policy Issues. In *Agricultural Development in the Third World,* Eicher C. and J. Staatz (eds), Baltimore: Johns Hopkins Press.

Coelli, T.J, 1995. Estimators and Hypothesis Tests for a Stochastic Frontier Function: A Monte Carlo Analysis. *Journal of Productivity Analysis* 6: 386-405.

Coelli, T. J. 1996. A Guide to FRONTIER Version 4.1: A Computer Program for Frontier Production Function Estimation. *CEPA Working Paper 96/07.* Center for Efficiency and Productivity Analysis, University of New England, Armidale.

Coelli, T.J., D.S. Rao and G.E. Battese. 1998. *An Introduction to Efficiency and Productivity Analysis*, Kluwert Academic Publishers.

Cohen, J. 1974. Rural Change in Ethiopia: The Chilalo Agricultural Development Unit, *Economic Development and Cultural Change*, No.22 (4).

Colman, D. and T. Young. 1990. *Principle of Agricultural Economics: Markets and Prices in Less Developed Countries*. Cambridge University Press.

Consultant Advisory Committee (CAC). 2004. Assuring Food and Nutrition Security in Africa by 2020: *A Way Forward from the 2020 Africa Conference*, Kampala, Uganda, IFPRI, *http://www.ifpri.org/2020Africaconference/index.htm.*

Cornwell, C., P. Schmidt and R. Sickles. 1990. Production Frontier with Cross-sectional and Time-series Variation in Efficiency Levels. *Journal of Econometrics, 46: 185-200.*

CSA. 2000. Central Statistical Authority Statistical Abstract, Addis Ababa, Ethiopia.

CSA. 2003. Central Statistical Authority Statistical Abstract. Addis Ababa, Ethiopia.

Curtiss, J. 2002. Efficiency and Structural Changes in Transition: A Stochastic Frontier Analysis of Czech Crop Production: A Ph.D. diss., Institutional Change in Agriculture and Natural Resources Vol. 12, Aachen: Shaker Verlag.

de Janvry, A., K. Nigel and E. Sadoulet. 1997. Agricultural and Rural Development Policy In Latin America: New Directions and New Challenges", *Working Paper* No. 815, University of California, Berkeley.

Dean, J.W. and M. Perlman. 1998. Harvey Lebenstein as a Pioneer over Time. *Journal of the Royal Economic Society* 108, Nr. 446: 132-152.

Deininger, K. 1995. Collective Agricultural Production: A Solution for Transition Economies? *World Development* 23 (8): 1317-1334.

Dejene, A. 1990. The Evolution of Rural Development Policies. In *Rural Development Options in Ethiopia,* Siegfried P., et al. (eds.), London: Zed Books Ltd.

Dejene, A. 1997. Transforming Peasant Agriculture: A Conceptual Framework. In *Ethiopian Agriculture: Problems Of Transformation: Proceedings Of The 4th Annual Conference on the Ethiopian Economy,* Dejene, A. and Mulat, D. (eds), Addis Ababa, Ethiopia.

Delgado, C.L., J. Hopkins and A.K. Valerie. 1994. *Agricultural Growth Linkages in Sub-Saharan Africa*, Washington, D.C., USAID.

Delgado, C.L. and J.W. Mellor. 1984. A Structural View of Policy Issues in African Agricultural Development. *American Journal of Agricultural Economics,* Vol. 66, 665-70.

Diewert, W. 1976. Exact and Superlative Index number. *Journal of Econometrics 4.*

Diewert, W. and T.J. Wales. 1987. Flexible Functional Forms and Global Curvature Condition. *Econometrica* 55(1): 43-68.

Dyckhoff, H. and K. Allen. 1997. Produktion und entischeidungstheoretische Begründung der Data Envelopment Analysis (DEA). Arbeitsbericht des Institutes für Wirtschaftswissenschaften der RWTH. Aechen.

Eldon, B., V. Christopher, B. Richard and A. Somwaru. 1997. Agricultural Productivity Revisited. *American Journal of Agriculture*, Vol. 79, no. 4

Eleni, G. M, and B. F. Hohnson. 1999. Accelerating Africa's Structural Transformation: Lessons from East Asia. Washington, D.C., USA, IFPRI.

Eleni, G. M, P. Dorosh and C. Barrett. 2002. Technological Change and Price in Agriculture: Conceptual and Comparative Perspectives, Washington, D.C., IFPRI,

Ellis, F. 1993. Private Trade and Public Role in Staple Food Marketing: The Case of Rice in Indonesia. *Food Policy* 1993: 428-438.

Engle, R. 1982. Autoregressive Conditional Heteroscedacity with Estimates of Variance of U.K. Inflation. *Econometrica* 50: 978-1008.

Engle, R.F. and C. W. Granger. 1987. Cointergartion and Error Correction: Representation, Estimation, and Testing. *Econometrica*, 1057-72.

EPRDF. 1995. Five-Year Development program, EPRDF publication, Addis Ababa.

EPRDF. 1996. Food security Strategy, Prepared for Consultative Group Meeting of December 10-12, Addis Ababa

Ethiopian Seed Enterprise (ESE). 2001. Seed Marketing Study, Final Report, Addis Ababa.

Fadani, A. 1999. Agricultural Price Policy and Export and Food Production in Cameroon: A farming Systems Analysis of Pricing Policies; The Case of Coffee-based Farming Systems, Ph.D. diss. University of Hohenheim, Frankfurt, Peter Lang.

Fahlbeck, L. 1996. *Essay in Transaction Cost Economics*. Uppsala: Swedish University of Agricultural Sciences.

Fan, S. 1991. Effect of Technological Change and Institutional Reform on Production Growth in Chinese Agriculture. *American Journal of Agriculture,* Vol. 73, No. 2: 266-275.

FAO. 2002. Food and Agriculture Organization Annual Report 2002, Addis Ababa.

FAO. 1991. The State of Food and Agriculture in 1990, Rome.

FAO/UNDP. 1984. Ethiopia: Land Use, Production Regions and Farming System Inventory. *Technical Report* 3, Vol 1, Rome, FAO.

Färe, R., S. Grosskopf and Lowell. 1985. *The Measurement of Efficiency in Production*. Boston, Dordrecht, London: Kluwer Academic Publisher.

Farrell, M.J, 1957. The measurement of Production Efficiency. *Journal of the Royal Statistical Society,* Series A-120.

Fei, J. C. H. and Ranis, G. 1964. *Development of the Labor-Surplus Economy*. Homewood, Richard D. Irwin: 55.

Fogel, R. 1994. Economic Growth, Population Theory and Physiology: The Bearing of Long-Term Processes on the Making of Economic Policy. *American Economic Review* 84, No. 3: 369-95.

Frank, A. G. 1976. *Capitalism and Underdevelopment in Latin America*. New York, Monthly Review Press.

Gavian, S. and S. Ehui. 1999. Measuring the Production Efficiency of Alternative Land Contracts in a Mixed Crop-Livestock System in Ethiopia. *Agricultural Economics* 20: 37-49.

Girma, B. 2002. The Role of Ethiopian Grain Trade Enterprise in Price Policy. A Paper Presented on EDI/IFPRI Workshop on the Diffusion of Agricultural Technology and Price Policy in Ethiopia, Draft, Addis Ababa.

Gittinger, J.P. 1982. *Economic Analysis of Agricultural Projects,* 2nd edition, USA, Maryland, Baltimore, The Johns Hopkins University Press.

Greene, W.H. 1993. *Econometric Analysis*, 2nd edition, New York University

Groenewald, J.A. 1991. Returns to Size and Structure of Agriculture: A Suggested Interpretation. *Development Southern Africa* 8(3): 329-342.

Haggblade, S., P, Hazell, and J. Brown. 1989. Farm-Non-Farm Linkages in Sub-Saharan Africa. *World Development* 17: 1173-1201.

Hamza, A. and T. Azanaw. 1995. Structural Adjustment Policy and Ethiopian Agriculture: An Assessment of Short Run Response and Structural Problems. In *Ethiopian Agriculture: Problems of Transformation: Proceedings of the 4th Annual Conference on the Ethiopian Economy*, Dejene A. and Mulat, D. (eds), Addis Ababa, Ethiopia

Harris, R. 1995. *Using Cointegration Analysis in Econometric Modeling*, London, Harvester Wheatsheaf.

Hayami, Y. and W. Ruttan. 1985. *Agricultural Development: An International Perspective, Revised and Expanded*, Baltimore, Johns University press.

Heidhues, F. 1999. The Social Dimension of Policy Reforms and World Bank Case Studies, Institute Für Agra-und Spzialwissenschaft in den Tropen und Subtropen, Universität Hohenheim.

Heidhues, F. and K. Beaktrice (eds.). 1992. Food and Agricultural Policies under Structural Adjustment: Seminar of the European Association of Agricultural Economists, Hohenheim University, Frankfurt, Peter Lang.

Heidhues, F. and Kamajiou (eds.). 1996. Agricultural Policy Analysis: Proceedings on International Seminar, University of Dschang, Cameroon,

Herman, S., Bezabi E., Berehanu A., Andrzei, B. and Shimelis, W/H. 1991. Farming Systems and Farm managment Practices of Smallholders in Harargeh Highlands, Wissenschaftsverlag Vauk Keil.

Hirschmann, A. O. 1958. *The Strategy of Economic Development*. New Haven CT. Yale University Press

Hoekman, B. and A. Kym. 2000. Developing-Country Agriculture and the New Trade Agenda. *Journal of Economic Development and Cultural Change*, Vol. 49, No.1, 171-80.

Husian, I. 1995. The Macroeconomics of Adjustment in Sub-Saharan Countries: Results and Lessons. In *Agricultural Perceptiveness: Market Forces and Policy Choice. Proceedings of the Twenty-Second International Conference of Agricultural Economics,* Peter, G.H and D. H. Douglas (eds.), Harare, Zimbabwe, U.K, Darmouth Publishing.

ICRAF. 1990. Agroforestry: Potentials, Research Needs for Ethiopian Highlands, Hoekstra, D., Torquebiau, E. and B. Badege (eds.), *Rapport AFRENA* Report, No. 21.

Jayne, T.S, Govereh, J. Mwanaumo, A., Chapoto, A, and Nyoro J.K. 2002. False Promise or False Premise? The Experience of Food and Input Market. In *Prospective on Agricultural Transformation: A Review from Africa*, USA, Nova Science Publisher,

Jayne, T.S., I. J. Minde and A. K. Gem (eds). 2002. Perspectives on Agricultural Transformation: A View from Africa, USA.

Johansen, S. 1988. Statistical Analysis of Cointegration Vectors. *Jounal of Economic Dynamics and Control*, 12: 231-254.

Johansen, S. 1991. *Statistical Analysis of Cointegration for I(2) Variables*, University of Copenhagen.

Johansen, S. 1993. Estimating Long-run Relationships in Economics: A Comparison of Different Approaches. *Journal of Econometrics*, 57, 53-68.

Johnson, D.G. 1973. *World Agriculture in Disarray*. Fontana. London Johsnton, 497-540. Ithaca: Cornell University Press.

Johnston, B. F. and P. Kilby. 1975. *Agriculture and Structural Transformation: Economic Strategy in Less Developing Countries*. London: Oxford University Press

Johnston, D., F. Bruce and W. Mellor. 1961. The Role of Agriculture in Economic Development. *American Economic Review* 51(4): 566-93.

Jorgenson, D. W. 1961. The Development of a Dual Economy. *Economic Journal* 71*: 309-334.*

Junanker, P.N. 1989. The Response of Peasant Farmers to Price Incentives: The Use and Misuse of Profit Function. *The Journal of Development Studies*, Vol, 25.

Kaldor, N. 1957. A Model of Economic Growth. *Economic Journal*, 68: 591-624.

Kalirajan, K. 1994. Intersectional Linkages in India: An Empirical Testing. Paper Presented in Indian Conference, Mysore, India.

Kalirajan, K., M. Obwona, and S. Zhao. 1996. Decomposition of Total Factor Productivity Growth: The Case of Chinese Agricultural Growth before and After Reforms. *American Journal of Agricultural Economics,* Vol. 78, No, 2

Kang, K. and V. Ramachandran. 1999. Economic Transformation in Korea: Rapid Growth without an agricultural Transformation? *Journal of Economic Development and Cultural Change*, Vol. 47, No.4: 783-802

Kirk, D. and B. Pillet. 1998. Fertility Levels. Trends, and Differentials in Sub-Saharan Africa in 1980s and 1990s. *Studies in family Planning*, Vol. 29, No. 1.

Kodhek, G., T.S. Jayne and I, Minde. 1999. African Perspectives on Progress and Challenges in Agricultural Transformation, Policy Synthesis for Cooperating USAID Offices and Country Missions, Global Bureau, Office of Agriculture and Food Security, *http:/www.aec.msu.edu.* Accessed in March 2003.

Koning, N. 2002. Should Africa Protect its Farmers to Revitalize its Economy? Gatekeeper Series No., 105, South Africa.

Krishna, R. 1967. Agricultural Price Policy and Economic Development. In *Agricultural Development and Economic Growth*, Herman M. Southworth and F. Bruce (eds.), USA.

Kuznets, S. 1966. *Modern Economic Growth*, New haven, Yale University Press.

Kuznets, S. 1973. Modern Economic Growth: Findings and Reflections. *America Economic Review*, 53, 246-270.

Lee, T. H. 1971. Intersectoral Caital Flows in Economic Development Success. Taiwan 1895-1960, Ithaca: Cornell University Press.

Leibenstein, H. 1966. Allocative Efficiency vs. X-Efficiency. *American Economic Review,* Vol. 56: 392-415.

Lelissa, C. 1998. The Determinants of Adoption, Intensity and Profitability of Fertilizer Use: The Case of Ejere District, West Showa Zone. M.Sc Thesis, Addis Ababa University.

Lewis, W. A. 1954. Economic Development with Unlimited Supply of Labor. *Manchester School of Economic and Social Studies* 22(2): 139-91

Lewis, W. A. 1955. *The Theory of Economic Growth*. London: George Allen and Unwin

Lewis, W. A. 1978. *The Evolution of the International Economic Order*. Pricneton, University Press.

Lewis, W. A. 1979. The *Dual Economy Revisited: Industrialization is Dependent up on Agricultural Improvement*. Manchester School 47: 211-29 (143).

Low, L. 2001. Globalization and Poverty Reduction: Can the Rural Poor Benefit from Globalization?: An Asian Perspective, Paper for Expert Group Meeting on Globalization and Poverty Reduction, Washington D.C., World Bank.

Lundvall, K. and G.E Battese. 1998. Firm Size, Age and Efficiency: Evidence from Kenyan Manufacturing Firms. *Center for Efficiency and Productivity Analysis*, Working Paper No. 7/98, University of New England, Australia.

Mackinnon, J. 1991. Critical Values for Cointegration Tests. In *Long run Economic Relationships,* Engle and Granger (eds.) Oxford University Press. PP. 267-76

Martin, W. and D. Mitra. 2001. Productivity Growth and Convergence in Agriculture versus Manufacturing. *Journal of Economics Development and Cultural Change*, Vol. 49, No. 2: 403-422.

Mathijs, E., A. Maertens and L. Vranken. 2001. Technical Efficiency and Farm Organization in Czech and Slovak Agriculture. Phare ACE project P97-8158-R, *Micro-economic Analysis of Farm Restructuring in Central and Eastern Europe*, Working Paper 3.

MEDaC. 1999. Survey of the Ethiopian Economy: Review of Post Reform Development (1992/93-1997/98), Addis Ababa, Ethiopia.

MEDaC. 2000. Survey of Road Construction in Ethiopia, Addis Ababa.

MEDaC. 2003. Annual Report, addis Ababa, Ethiopia.

Meerma, J. 1997. Reforming Agriculture: The World Bank Goes to Market. *A World Bank Operation Evaluation Study*. Washington, D.C, World Bank.

Mellor, J. W. 1962. Increasing Agricultural Production in Early Stage of Economic Development, Relations, Problems and Prospects. *Indian Journal of Agricultural Economics*, Vol. 27, No. 2 29-46.

Mellor, J. W. 1998. Agriculture on the Road to Industrialization. In *International Agricultural Development,* Eicher C.K and J.M. Staatz (eds), the Johns Hopkings University Press, Baltimore and London

Mellor, J.W. 1995. (ed). *Agriculture on the Road to Industrialization*, Baltimore: John Hopkins University Press.

Mellor, J. W. 1976. The New Economics of Growth: Strategy for Indian and Developing World. Ithaca NY: Cornell University Press.

Mellor, J. W. and L. Uma, 1973. Growth Linkages of the New Food Grain Technologies. *Indian Journal of Agricultural Economics* 28 (1): 35-55.

Mellor, J.W. 1985. Determinants of Rural Poverty: The dynamics of Production, Technology and Price. In *Agricultural Change and Rural Poverty,* Mellor, J.W. and G.M. Desai (eds.), Johns Hopkins University Press.

Moa, Y. 1992. Institutional Organizations in Taiwan's Agricultural Development, In *Rural Development in Taiwan and Mainland China,* Peter Calkins, Wen Chern, Francis Tuan (eds), Boulder: Westview Press.

MOA. 1984. Highland Reclamation Study, Working Paper no. 24.

Mohammed, H., F. Hussen, W. Mwangi, and Belay, K. 2000. Farctors Influencing Technical Efficiency of Barley Production in Asassa District of Southeastern Ethiopia. *Ethiopian Journal of Agricultural Economics,* Vol. 4, No. 1and 2.

Moller, J. W. and F. B. Johnston. 1984. The World Food Equation: Interrelations Among Development, Employment and Food Consumption. *Journal of Economics Literature*, Vol. 22.

Moller, J. W. and S. Mohinder. 1992. Agriculture in Economic Development: Theories, Findings and Challenges in Asian Context. In *A Survey of Agricultural Economics Literature* Vol. 4, University of Minnesota Press.

Morduch, J. 1991. Risks and Welfare in Developing Countries. Ph.D. diss. in Harvard University.

Mosher, A.T. 1966. *Getting Agriculture Moving: Essentials For Development and Modernization*: New York, Praeger

Mosley, P. and S. Lawrence. 1989. Structural Adjustment and Agricultural Performance in Sub-Saharan Africa 1980-87. *Journal of International Development,* Vol. 1, No. 3, 321-355.

MPED. 1993. An Economic Development Strategy for Ethiopia: A Comprehensive Guidelines and A Development Strategy for Future, Ministry of Planning and Economic Development, Addis Ababa.

Mulat, D. 1999. The Challenge of Increasing Food Production in Ethiopia. In *The Ethiopian Economy Performance and Evaluation: Proceedings of the Eighth Annual Conference of the Ethiopian Economy*, Alemayehu, G. and Berhanu, N. (eds.), Addis Ababa, Ethiopian Economic Association

Murova, O., M. A. Treueblood and K.H. Coble. 2000. Efficiency Analysis of Ukrainian Agriculture. Paper Presented at the North American Productivity Workshop, Schenactacdy, NY.

Nabseth, L. and G.F. Ray (eds). 1974. *The Diffusion of Industrial Processes*, Cambridge University Press, 148-9.

Narayana, N., K.S. Parikh and T.N. Srinivasan. 1991. Agriculture, Growth and Redistribution of Income: *Food Policy Analysis*, , the Netherlands, Kluwer Academic Publisher

National Bank of Ethiopia. 2003. National Bank of Ethiopia Annual Report, Addis Ababa.

Nelson, D. and C.Q. Coa. 1992. Inequality Constraints in the Univariate GARCH Model. *Journal of Business and Economics Statistics* 10: 49-55

Nerlove, M. 1958. *The Dynamics of Supply: Estimation of Farmers' Response to Price*. Baltimore: Johns Hopkins University Press

Newbery, D. M. and J. E. Stiglitz,. 1981. *The Theory of Commodity Price Stabilization*. Oxford: Oxford University Press.

Nicholls, W. 1964. The Place of Agriculture in Economic Development. *Agriculture in Economic Development,* Cark K. and Lawrence W. (eds.), New York

Nurkse, R. 1976. The Case of Balanced Growth. In *Leading Issues in Economic Development*, Meier, G. M (ed.), Oxford University Press, 640-643.

OECOD. 1983. Historical Statistics 1960-1981. OECOD, Paris.

Ohkawa, K. 1964. Concurrent Growth of Agriculture with Industry: A Study of Japanese Case. In *International Explorations of Agricultural Economic*, Roger N. Dixey (ed.), 201-12. Ames, Iowa State University Press.

Ohkawa, K., and F. Bruce. 1969. The Transferability of Japanese Pattern of Mdernizing Traditional Agriculture. In *The Role of Agriculture in Economic Development*, Erik Thorbecke (eds.), 277-303. New York, National Bureau of Economic Research.

Okey, Y. 2004. Governance in Africa: *http://www.irinnews.org/report.asp?ReportID=37239&SelectRegion=Horn_of_Africa&SelectCountry=AFRICA.* Accessed in Oktober 2004.

Oxfam. 1993. *Africa, Make or Break: Actions For Recovery*: Oxford: Oxfam

Parikh, K.S., G. Fisher, K. Frohberg and O. Gulbrandsen. 1988. Towards Free Trade in Agriculture, Martinus, Dordrecht, The Netherlands

Parikh, K.S. 1992. The Importance of Food and Agricultural Policies in the Success of Structural Reform in India. In *Food Policy and Agricultural Polices Under Structural Adjustment-Seminar of the European Association of Agricultural Economics,* Heidhues, F. and B. Knerr (eds.), Germany, Peter Lang.

Park, A. and J. Bruce 1995. Rural Development and Dynamic Externalities in Taiwan's Structural Transformation. *Journal of Economic Development and Cultural Change*, Vol. 4, No.1: 181-208.

Piesse, J. and C. Thirtle. 2000. A Stochastic Frontier Approach to Firm Level Efficiency, Technological Change and Productivity during Early Transition in Hungary. *Journal of Comparative Economics* 28 (3): 473-501.

Pitt, M. and L. Lee. 1981. The Measurement and Sources of Technical Inefficiency in the Indonesian Weaving Industry. *Journal of Development Economics* 9: 43-64.

Porter, P.K and G.W. Schully. 1987. Economic Efficiency in Cooperatives. *Journal of Law and economics*, XXX: 489-512.

Prebish, R. 1950. *The Economic Development of Latin America and its Principal Problems*, U.N (222), New York

Pyakurel, B. 2002. Nepal's Agriculture Development: A Note on Failed Promise of Rural Transformation, *http://www.plusmedium.com*. Accessed in March 2003.

Rangarajan, C. 1982. Agricultural Growth and Industrial Performance in India, *Research Report* 33, International Food Policy Research Institute.

Rao, P.N. and C. H. Hanumantha. 1994. Agricultural Growth, Rural Poverty and Environmental Degradation in India. Delhi: Oxford University Press.

Rayner, A., and S. J. Welham. 1995. Economic and Statistical considerations in the measurement of Total Factor Productivity (TFP), in Agricultural Productivity, Banett Vic, Page Reg and Steiner Roy (eds).

Roche, M. and K. Mcquinn. 2003. Grain Price Volatility in a Small Open Economy. *JEL classification*: C32, F31, G25, Q14

Rostow, W.W. 1960. *The State of Economic Growth*, Cambridge University Press.

Rostow, W.W. 1975. *How It All Began*, Methuen, London,.

Ruttan, V. W. and Y. Hayami. 1998. Induced Innovation Model of Agricultural Development. In *International Agricultural Development,* Eicher C.K and J.M. Staatz (eds.), Baltimore and London, Johns Hopkings University Press.

Ruttan, V.W. 1998. Models of Agricultural Development. In *International Agricultural Development,* Eicher C.K and J. M. Staatz (eds), Baltimore and London, the Johns Hopkings University Press.

Sadoulet, E. and A. de Janvry. 1995. *Quantitative Development Policy Analyses*, Baltimore and London Johns Hopkins University Press.

Salim, R. and K. P. Kalirajan. 1999. Sources of Output Growth in Bangladesh Food Processing Industries: A Decomposition Analysis. *The Development Economics*. XXXVII: 355-374.

Schiff, M. 1997. Aggregate Agricultural Supply Response in Developing Countries: A Survey of Selected Issues. *Journal of Economic Development and Cultural Change*; Vol. 45, No.2, 393-409.

Schlüter, A. 2001. Institutioneller Wandel und Transfortion- Restitution. Trasnformation und Privatisierung in der tschechischen Landwirtschaft. Institutioneller Wandel der Landwirtschaft und Resourcennutzung (3). Aachen: Shaker Verlag.

Schmitt, G. 1991. The Organization of Farm Production and its Implications for Agricultural Economics and Policy. Paper Presented at the Policy Article Prize Seminar, Department of Agricultural and Applied Economics, University of Minnesota.

Schuh, G. E. 2002. Globalization and Rural Development. Paper presented a Conference on Rural Community Vitality in a Global Economy, Humphrey Institute of Public Affairs, University of Minnesota, Minneapolis, 2002.

Schultz, T. W. 1964. *Transforming Traditional Agriculture*, Yale University Press, New Haven.

Schultz, T.W. 1978. *Distortion of Agricultural Incentives*, Bloomington.

Senait, R. 2002. The economics of Managing Land Resources towards Sustainability in the Highlands of Ethiopia. Farming and Rural Systems Economics; PhD. Dissertation, University of Hohenheim,Weikersheim.

Seyoum, E., G. Battese and E.M. Fleming. 1998. Technical Efficiency and Productivity of Maize Producers in Ethiopia: A Study of Farmers within and Outside Sasakawa-Global 2000 Project. *Journal of Agricultural Economics,* 19: 341-348.

Shank, R. 1996. The 1996 Fertilizer Situation: Progress, Problems and Programs, UNDP Emergency Unit for Ethiopia, United Nations Development Program.

Sicular, T. 1999. Investments Vs Institutional Transformation to Promote Growth: Perspectives from China, http://*www.econ.yale.edu*. Accessed in January 2004.

Sotnikov, S. 1998. Evaluating the Effect of Price and Trade Liberalization on the Technical Efficiency of Agricutlural Production in a Transition Economy: The Case of Russia. *European Review of Agricultural Economics* 25 (3): 412-31.

Staatz, J.M. 1998. What is Agricultural Transformation? *http://www.aec.msu.edu*, USA, Michigan State University.

Staatz, J.M. and C.K. Eicher. 1998. Agricultural Development Ideas and Historical Perspectives. In *International Agricultural Developmet*, Eicker, C.K and Staatz, J.M (eds.), Baltimore and London, The Johns Hopkins University Press.

Stiglitz Joseph. 2002. Globalization and Its Discontents. *htt/www.thirdworldtraveler.com/Globalisms_Discontents.html.* Accessed in September 2004.

Straus, J. 1986. Does Better Nutrition Raise Farm Productivity? *Journal of Political Economy* 94, No. 2: 297-320.

Sundrum, R. M. 1983. *Development Economics: A Framework fir Analysis and Policy*. New York, John Wiley and Sons.

Tegegne, G/E. 1995. An Assessment of Ethiopia's Agricultural Resources. In *Ethiopian Agriculture: Problems of Transformation: Proceedings of The 4th Annual Conference on the Ethiopian Economy*, Dejene, A. and Mulat, D. (eds.), Addis Ababa, Ethiopian Economic Society.

Tennassie, N. 1985. Agricultural Research and Extension in Ethiopia: The State of the Art. Ethiopia, IAR.

TGE. 1991. Ethiopia Economic Policy during the Transitional Period, Addis Ababa Ethiopia

Thirtle, C. and W, Ruttan. 1987. *The Role of Demand and Supply in the Generation and Diffusion of Technical Change*. Harwood Academic Publisher.

Timmer, P. 1986. *Getting Prices Right: The Scope and Limits of Agricultural Price Policy*. Ithaca: Cornell University Press.

Timmer, P. 1988. The Agricultural Transformation. In *Hand Book of Development Economics,* Chenery H. and T.N. Srnivasan (eds.), Vol. 1, Elsevier Science.

Timmer, P. 1989. Food Price Policy: The Rational for Government Intervention. *Food Policy* 14, No. 1: 17-42.

Timmer, P. 1992. Agriculture and Development Revised. *Agricultural Systems* 38, No. 5:1-35.

Timmer, P. 1993. *Why Market and Politics Undervalue the Role of Agriculture in Economic Development.* Benjamin H. Hibbard Memorial Lecture Series, Department of Agricultural Economics, University of Wisconsin, Madison.

Timmer, P. 1995. Getting Agriculture Moving: Do Markets Provide the Right Signals? *Food Policy* 20, No. 5: 455-72.

Timmer, P. 1998. The Agricultural Transformation. In *International Agricultural Development*, 3rd, Eicher C.K. and J. M. Staatz (eds), The Johns Hopkins University press.

Todaro, M. 1989. *Economic development in the Third World*, 4th edition, New York, Longman.

Todaro, M. and C. S. Smith. 2003. *Economic development*, 8th edition, New York, Longman Boston, Pearson.

Turner, M. 1998. Agricultural Revolution, *http://www.eh.net*. Accessed in September 2003.

Tybout, J. and M. Daniel. 1995. Trade Liberalization and Dimensions of Efficiency Changes in Mexican Manufacturing Industries. *Journal of International Economics*, 39(1/2): 53-78.

Voigt, P. and V. Uvarovsky. 2001. Development in Productivity in Efficiency in Russia's Agriculture: The Transition Period. *Quarterly Journal of International Agriculture* 40, No. 1: 445-66.

Von Braun, J., F. Bellin-Sesay, T. Feldbrügge and F. Heidhues. 1998. *Verbesserung der Ernährung in Entiwicklungsländern: Strategien und Politikempfehlungen*. Germany, Weltforum Verlag.

Von Oppen, M., B.K. Njehia and A. Ijaimi. 1997. The Impact of Market Access on Agricultural Productivity: Lessons from India, Kenya and the Sudan. *Journal of Development:* Vol. 9, No. 1.

Wade, R. 1990. *Governing the Market: Economic Theory and the Role of Government in East Asian Industrialization*, Princeton, Princeton University Press.

Wolday, A. 2002. The Structuring and Functioning of the Post PADETS Grain Marketing System in Ethiopia, draft, A Paper Presented on EDRI/EFPRI Workshop on Diffusion of Agricultural Technology and Price Policy in Ethiopia (Draft), Ethiopia

Woodldridge, J. M. 2003. *Introductory Econometrics: A Modern Approach*, 2nd edition, U.S.A, Thomson, South-Western.

World Bank and IMF. 2004. Overview of Poverty Reduction Strategies. *http://www.worldbank.org/poverty/strategies/overview.htm*. Accessed in August 2004.

World Bank Group. 2001. Comprehensive Development Framework: PRSP/CDF, *http://web.worldbank.org/external/projects/strategies/cdf.* Accessed in August 2004.

World Bank. 1982. *World Development Report 1982*. New York: Oxford University Press

World Bank. 1986. *World Development Report 1986*. New York: Oxford University Press.

World Bank. 1994. Ethiopia: National Fertilizer Project Staff Appraisal Report, (Draft).

Zeller, M., G. Buchenschrieder, J. von Braun and F. Heidhues. 1997. Rural Finance for Food Security for the Poor: Implications for Research and Policy, *Policy Review* 4, Washington, D.C., IFPRI.

Annexes

Table A3.1: Granger Causality Test for Sectoral Growth Relationship

The Null Hypothesis	Lag Structure	
	One period	Two periods
Log industry does not Granger Cause Log agriculture	2.82 (0.13)	1.56 (0.22)
Log agriculture does not Granger Cause Log industry	0.43 (0.52)	2.13 (0.13)
Log service does not Granger Cause Log agriculture	8.46 (0.006)*	5.33 (0.009)*
Log agriculture does not Granger Cause Log service	0.63 (0.43)	0.41(0.67)
Log service does not Granger Cause Log industry	1.55 (0.22)	3.98 (0.03)*
Log industry does not Granger Cause Log service	0.57 (0.45)	2.21 (0.13)

Source: Own computation based on data from CSA and National Bank series of reports

Notes: Figures in parenthesis are *P*-values for the F-test; the alternative hypothesis is rejected at conventional level of significance in each of the cases (indicated by *). Accordingly, the Granger Causality test shows that no significant causal relationship exists between the growth in the agricultural and industrial sectors; only the growth in the service sector was found to granger cause the growths in the other two sectors, but not vise versa.

Figure A3.1: Real GDP per Capita at 1981 Constant Factor Cost

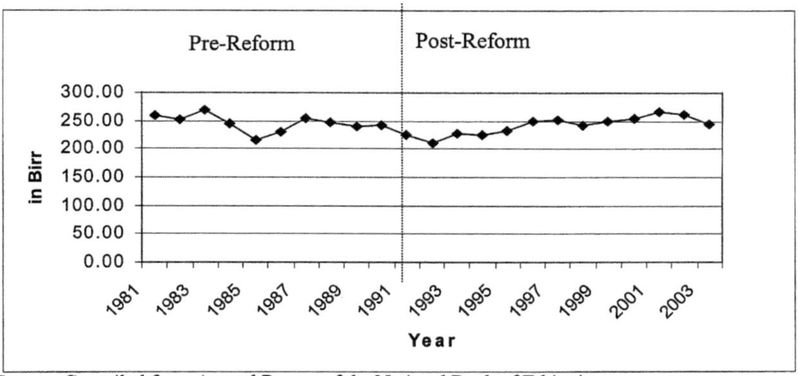

Source: Compiled from Annual Report of the National Bank of Ethiopia.

Table A3.2: Global Comparison of Trends in Yields of Major Crops

Yield (100Kg/ha)	Year												
	1980	1985	1990	1994	1995	1996	1997	1998	1999	2000	2001	2002	2003
Cereals													
Africa South of Sahara	9.2	9.8	9.6	9.1	9.5	10.3	9.6	10.0	10.2	10.0	9.9	9.6	9.9
Asia Developing	20.5	24.5	27.8	29.4	30.2	31.4	31.4	32.2	32.8	31.9	32.5	32.5	32.5
Ethiopia	-	-	-	9.7	10.3	12.1	12.6	11.4	11.2	11.2	12.0	13.5	13.3
Germany	42.3	50.7	54.1	58.3	61.1	62.8	64.8	63.4	67.0	64.5	70.5	62.5	57.6
United States of America	37.7	47.6	47.5	55.6	46.4	51.8	52.8	56.8	57.3	58.5	58.9	55.7	60.3
Pulses.Total													
Africa South of Sahara	5.4	4.7	4.9	4.6	4.5	4.6	4.4	4.7	4.6	4.8	5.0	4.8	4.8
Asia Developing	5.8	6.7	7.2	7.3	7.5	7.1	7.2	7.2	7.3	7.4	7.0	7.6	7.1
Ethiopia	-	-	-	6.1	8.3	9.0	9.7	7.7	7.8	8.8	8.2	9.1	9.1
Germany	17.7	25.1	23.5	33.8	33.7	34.9	33.9	35.0	37.6	29.6	34.7	28.5	28.9
United States of America	16.2	16.3	17.4	17.8	19.2	17.4	19.4	18.2	19.8	18.8	18.3	18.9	18.0
Maize													
Africa South of Sahara	11.0	12.3	11.7	11.9	12.1	13.4	12.2	12.4	13.3	12.7	12.5	11.9	12.0
Asia Developing	23.3	26.3	33.1	34.7	36.1	38.6	34.0	39.9	38.3	35.7	37.2	38.4	38.6
Ethiopia				11.2	13.6	16.8	17.4	16.2	17.2	16.2	17.4	18.7	18.8
Germany	56.4	66.5	67.8	70.8	73.7	78.3	86.3	81.6	87.9	92.1	88.4	93.8	72.2
United States of America	57.1	74.1	74.4	87.0	71.2	79.8	79.5	84.4	84.0	85.9	86.7	81.6	89.2
Wheat													
Africa South of Sahara	13.5	13.8	16.4	14.0	14.8	15.5	16.5	17.3	14.1	14.5	16.2	15.7	15.9
Asia Developing	16.1	21.3	23.8	24.8	26.2	26.7	28.6	27.3	28.0	28.0	27.8	28.0	27.7
Ethiopia	-	-	-	10.7	13.1	12.1	12.9	13.7	11.1	11.6	13.3	14.0	14.0
Germany	47.4	58.3	62.7	67.6	68.9	72.9	72.7	72.0	75.4	72.8	78.8	69.1	65.0
United States of America	22.5	25.2	26.6	25.3	24.1	24.4	26.6	29.0	28.7	28.3	27.1	23.7	29.7

Source: FAO database

Table A3.3: Food Imports and Victims of Food Shortage

Year	Purchased food imports (000 MT)	Food aid Distributed (000 MT)	Drought- (Disaster-) Afflicted Population (million)	Proportion of Population Affected (%)
1980/81	124.22	239	2.82	7.7
1981/82	178.26	201	3.70	9.7
1982/83	259.47	368	3..30	8.5
1983/84	298.53	188	4.21	10.5
1984/85	601.9	1272	6.99	17.0
1985/86	846.78	926	6.14	14.5
1986/87	599.9	277	2.53	5.8
1987/88	565.62	1096	4.16	9.3
1988/89	408.9	461	5.35	11.6
1989/90	128.1	657	3.21	6.8
1990/91	401.52	925	7.22	14.8
1991/92	12.83	840	7.85	15.6
1992/93	404.6	519	4.97	9.6
1993/94	411.2	980	6.70	12.6
1994/95	622.47	683	3.99	7.3
1995/96	322.28	334	2.78	4.9
1996/97	13.1	206	3.36	5.8
1997/98	30.84	417	4.10	6.8
1998/99	225.03	511	7.19	11.7
1999/00	391.28	979	10.56	16.6
2000/01	420.44	983	4.6	7.10
2001/02	738.1	409	13.20	19.70
2002/03@	645.14	na	14.3	22.0

Source: Disaster Prevention and Preparedness Commission of Ethiopia. 2003.
@ denotes approximation; na -is information not available.

Table A4.1: Multiple Comparisons of Livestock and Oxen Power Changes over Years

Dependent Variable	(I) Year	(J) Year	Mean Difference (I-J) Bonferroni (I-J)
Total livestock unit owned	1994	1997	0.867
		1999	1.145*
		2001	1.687*
	1997	1994	-0.867
		1999	0.278
		2001	0.812*
	1999	1994	-1.145*
		1997	-0.278
		2001	0.531
Number of oxen and bull owned	1994	1997	-0.064
		1999	-0.279
		2001	-.409*
	1997	1994	0.006
		1999	-0.216
		2001	-0.356*
	1999	1994	0.279
		1997	0.210
		2001	-0.132

Source: Computed from the household survey
* The mean difference is significant at least at the 5% confidence level.

Table A4.2: Multiple Comparisons of Livestock and Oxen among locations

Dependent Variable	(I) Woreda	(J) Woreda	Mean Difference (I-J) Bonferroni
Total livestock owned	Ankober-Dinqi	Debrebrehan	-6.09*
		Shashemene	-1.59*
	Debrebrehan	Ankober-Dinqi	6.09*
		Shashemene	4.50*
	Shashemene	Ankober-Dinqi	1.59*
		Debrebrehan	-4.50*
Number of oxen/ bulls own	Ankober-Dinqi	Debrebrehan	-1.24*
		Shashemene	-0.24
	Debrebrehan	Ankober-Dinqi	1.24*
		Shashemene	1.00*
	Shashemene	Ankober-Dinqi	0.24
		Debrebrehan	-1.00*

Source: Computed based on the household survey
* The mean difference is significant at least at the 5% confidence level. Note that the Bonferron uses t tests to perform pairwise comparisons between group means. and controls overall error rate by setting the error rate for each test to the experimentwise error rate divided by the total number of tests. Hence. the observed significance level is adjusted for the fact that multiple comparisons are being made.

Table A4.3: Changes and Distributions of Oxen/bulls Owned by Location and Time

Location	Number of oxen/bulls owned	Year			
		1994	1997	1999	2001
Ankober-Dinqi	N	87	85	74	50
	zero	59.8	43.5	31.1	30.1
	one	26.4	24.7	21.6	22.0
	two	12.6	24.7	35.1	34.4
	more than two	1.1	7.1	12.2	13.5
Debrebrehan	N	183	171	172	62
	zero	19.1	11.1	12.8	8.1
	one	20.8	10.5	14.5	22.6
	two	20.2	40.9	40.1	37.1
	more than two	39.9	37.4	32.6	32.3
Shashamane	N	102	101	92	56
	zero	39.20	36.6	26.1	24.40
	one	34.3	33.7	31.5	32.80
	two	10.8	20.8	27.2	25.50
	more than two	15.7	8.9	15.2	17.30

Source: Computed from the household survey
Note: na is not available

Table A4.4: Percentage of Samples Farmers Using Improved Seeds and Fertilizer and the Intensity of Application (kg/ha)

Location	Crop	1994	1997	1999	2001
Chmical Fertlizer:					
Ankober-Dinqi	Maize	13 (na)	24 (na)	32 (na)	36 (40)
	Tef	19 (na)	26 (60)	37 (110)	39(105)
Debrebrehan	Wheat	29 (na)	57 (61)	71 (74)	55 (64)
	Barley	45 (na)	48 (66)	82 (54)	98 (81)
Shashamane	Wheat	43 (na)	96 (86)	94 (125)	83 (92)
	Maize	47 (na)	51 (64)	86 (82)	74 (122)
	Tef	49 (na)	62 (72)	94 (101)	65 (120)
Improved Seed:					
Ankober-Dinqi	Maize	na	na	5.0	8.0
Shashamane	Maize	17.0	22.0	31.0	40.0
Debrebrehan	Wheat	na	4.5	12.0	18.0
Shashamane	Wheat	3.0	9.0	25.0	37.0

Source: Computed from the household survey
Note: Figures in the parenthesis are intensity of chemical fertilizer used (Kg/ha); na denotes information not available.

Table A4.5: Descriptive Statistics of Inputs and Output Variables Entered in the Tef Equation

Variables	Location	1994	1997	1999	2001
Area cultivated (ha)	Ankober-Dinqi	0.542	0.548	0.481	0.490
	Shashamane	0.304	0.329	0.327	0.415
	Pooled Mean	0.415	0.451	0.423	0.432
	Pooled SD	0.343	0.502	0.218	0.481
Labor equivalent	Ankober-Dinqi	2.486	2.819	2.381	2.54
	Shashamane	3.656	3.74	3.69	3.57
	Pooled Mean	3.11	3.65	2.87	3.15
	Pooled SD	1.66	1.94	1.57	2.97
Fertilizer used (kg)	Ankober-Dinqi	6.91	32.50	52.91	45.52
	Shashamane	25.22	22.00	33.48	49.80
	Pooled Mean	16.65	29.25	45.24	48.80
	Pooled SD	29.07	17.61	27.76	70.73
Number of oxen/bull owned	Ankober-Dinqi	0.727	1.476	1.692	1.65
	Shashamane	1.36	1.61	1.90	1.83
	Pooled Mean	1.06	1.52	1.77	1.83
	Pooled SD	1.15	0.89	1.31	1.64
Actual output (kg)	Ankober-Dinqi	133.18	181.55	287.69	204
	Shashamane	147.40	151.96	217.42	264
	Pooled Mean	140.74	171.08	267.71	2340
	Pooled SD	93.38	87.76	175.40	143
Yield (kg/ha)	Ankober-Dinqi	246	331	598	553
	Shashamane	485	462	665	521
	Pooled Mean	339	379	633	636

Source: Computed from the household survey
SD is Standard of Deviation.

Table A4.6: Descriptive Statistics of the Inputs and Output Variables Entered in the Wheat Equation

Variables	Location	1994	1997	1999	2001
Area cultivated (ha)	Debrebrehan	0.52	0.61	0.48	0.58
	Shashamane	0.47	0.86	0.49	0.62
	Pooled Mean	0.50	0.65	0.48	0.60
	Pooled SD	0.34	0.47	0.29	0.39
Labor equivalent	Debrebrehan	2.98	3.09	2.77	2.93
	Shahshamane	3.82	5.06	3.50	4.10
	Pooled Mean	3.30	3.42	3.01	3.52
	Pooled SD	1.54	1.65	1.49	1.84
Fertilizer used (kg)	Debrebrehan	29.95	23.99	41.12	56.78
	Shashamane	30.33	82.80	51.90	91.55
	Pooled Mean	30.09	33.85	44.73	68.43
	Pooled SD	50.62	43.87	43.84	48.83
Number of oxen/bull owned	Debrebrehan	2.46	2.52	2.51	2.38
	Shashamane	1.17	1.32	1.63	2.66
	Pooled Mean	1.97	2.32	2.21	2.42
	Pooled SD	1.78	1.69	1.40	1.48
Actual output (kg)	Debrebrehan	142.92	245.92	359.00	493.52
	Shashamane	626.92	998.00	698.26	842.81
	Pooled Mean	327.98	422.44	472.66	529.50
	Pooled SD	330.96	582.18	388.69	309.72
Yield (kg/ha)	Debrebrehan	275	403	748	851
	Shashamane	1334	1160	1425	1359
	Pooled Mean	656	650	985	883

Source: Computed from the household survey

Table A4.7: Descriptive Statistics of the Inputs and Output Variables Entered in the Maize Equation

Variables	Location	1994	1997	1999	2001
Area cultivated (ha)	Ankober-Dinqi	0.393	0.560	0.292	0.471
	Shashamane	0.300	0.384	0.354	0.362
	Pooled Mean	0.322	0.440	0.333	0.401
	Pooled SD	0.199	0.369	0.197	0.253
Labor equivalent	Ankober-Dinqi	2.200	3.200	2.706	2.85
	Shashamane	4.265	4.390	3.417	3.35
	Pooled Mean	3.78	3.97	3.16	3.05
	Pooled SD	1.79	2.10	1.64	2.30
Fertilizer used (kg)	Ankober-Dinqi	3.57	5.42	10.47	17.80
	Shashamane	45.14	22.51	36.43	41.61
	Pooled Mean	34.61	15.31	24.54	32.61
	Pooled SD	43.02	33.39	35.51	29.57
Number of oxen/bull owned	Ankober-Dinqi	0.57	1.54	1.82	1.92
	Shashamane	1.96	1.12	1.61	1.83
	Pooled Mean	1.63	1.25	1.68	1.84
	Pooled SD	1.47	0.97	1.29	1.57
Actual output (kg)	Ankober-Dinqi	105.71	129.58	165.15	172.30
	Shashamane	273.91	666.67	736.36	723.00
	Pooled Mean	234.67	494.80	542.15	723.00
	Pooled SD	169.52	623.11	629.02	253.98
Yield (kg/ha)	Ankober-Dinqi	269	231	566	603
	Shashamane	913	1736	2080	2499
	Pooled	729	1125	1628	2002

Source: Computed from the household survey

Table A4.8: Descriptive Statistics of Farmer Specific Technical Efficiency Level in Tef. Wheat and Maize Production

Year	Locations	Minimum	Maximum	Mean	SD
Tef Farmers					
	Ankober-Dinqi	0.28	0.82	0.731	0.160
1994	Shashamane	0.22	0.86	0.790	0.134
	Pooled	0.22	0.86	0.761	0.148
	Ankober-Dinqi	0.57	0.93	0.831	0.099
1997	Shashamane	0.83	0.95	0.879	0.038
	Pooled	0.57	0.95	0.849	0.084
	Ankober-Dinqi	0.82	0.97	0.932	0.043
1999	Shashamane	0.93	0.98	0.955	0.015
	Pooled	0.82	0.98	0.941	0.036
2001	Shashamane	0.78	0.99	0.947	0.033
	Ankober	0.77	0.98	0.952	0.031
	Pooled	0.78	0.99	0.948	0.411
Wheat Farmers					
	Debrebrehan	0.37	0.88	0.77	0.09
1994	Shashamane	0.67	0.85	0.78	0.05
	Pooled Mean	0.37	0.88	0.77	0.08
	Debrebrehan	0.54	0.93	0.85	0.07
1997	Shashamane	0.78	0.90	0.85	0.04
	Pooled Mean	0.54	0.93	0.85	0.06
	Debrebrehan	0.68	0.95	0.90	0.05
1999	Shashamane	0.85	0.94	0.91	0.03
	Pooled Mean	0.68	0.95	0.90	0.04
	Debrebrehan	0.79	0.97	0.93	0.03
2001	Shashamane	0.89	0.97	0.95	0.02
	Pooled Mean	0.79	0.97	0.94	0.03
Maize Farmers					
	Ankober-Dinqi	0.81	0.94	0.87	0.092
1994	Shashamane	0.81	0.95	0.89	0.131
	Pooled average	0.81	0.95	0.88	0.112
	Ankober-Dinqi	0.71	0.92	0.81	0.124
1997	Shashamane	0.69	0.91	0.84	0.189
	Pooled average	0.69	0.92	0.82	0.163
	Ankober-Dinqi	0.55	0.87	0.83	0.146
1999	Shashamane	0.56	0.84	0.77	0.255
	Pooled average	0.55	0.87	0.79	0.183
	Shashamane	0.55	0.87	0.73	0.215
2001	Ankober_Dinqi	0.52	0.85	0.79	0.253
	Pooled Mean	0.54	0.86	0.75	0.242

Source: Computed from the household survey

Table A4.9: Technical Efficiency. Technological change and TFP by Crop and Location

Crop	Location	Year	TE	TCH	TFP
Tef					
		1994	1.000	1.000	1.000
	Ankober-Dinqi	1997	1.137	0.592	0.673
		1999	1.275	0.805	1.026
		2001	1.295	0.807	1.045
		1994	1.000	1.000	1.000
	Shashamane	1997	1.113	0.379	0.422
		1999	1.210	0.554	0.670
		2001	1.199	0.641	0.769
Wheat					
		1994	1.000	1.000	1.000
	Debrebrehan	1997	1.104	1.377	1.520
		1999	1.170	1.343	1.571
		2001	1.208	1.325	1.601
		1994	1.000	1.000	1.000
	Shashamane	1997	1.101	1.366	1.503
		1999	1.167	1.331	1.553
		2001	1.218	1.310	1.596
Maize					
		1994	1.000	1.000	1.000
	Ankober-Dinqi	1997	0.931	1.015	0.945
		1999	0.954	1.202	1.147
		2001	0.910	1.220	1.110
		1994	1.000	1.000	1.000
	Shashamane	1997	0.944	1.294	1.222
		1999	0.865	1.514	1.310
		2001	0.820	1.702	1.400

Source: Computed from the household survey

Table: A6.1: Descriptive Statistics of Variables Used in the Analysis of VECM (1980-2003)

Description	Fertilizer price	Improved seed price	Cereal price	Cereal-fertilizer price ratio	Maize price	Maize-fertilizer price ratio	Tef price	Tef-fertilizer price ratio	Wheat price	Wheat-fertilizer price ratio
Mean	148.55	219	121.16	0.79	76.79	0.50	151.89	0.96	114.21	0.74
Maximum	256.20	284	206.67	1.55	147.00	0.75	264.00	1.29	209.00	1.04
Minimum	78.31	61	37.33	0.36	24.00	0.20	49.00	0.53	39.00	0.31
Std dev.	73.93	66	57.75	0.26	40.66	0.16	74.07	0.22	56.66	0.22
Skewness	0.40	-	0.13	1.09	0.46	-0.32	0.22	-0.48	0.41	-0.37
Kurtosis	1.35	-	1.49	4.83	1.74	2.40	1.54	2.20	1.84	1.82
CV	**49.77**	**23.07**	**47.67**	**33.55**	**52.95**	**31.45**	**48.76**	**22.68**	**49.61**	**30.18**
Jarque-Bera	3.25	-	2.04	6.79	1.92	0.61	1.84	1.23	1.59	1.53
Probability	0.20	-	0.36	0.03	0.38	0.74	0.40	0.54	0.45	0.46

Description	Cereal area (000ha)	Cereal output (000MT)	Cereal yield (ton/ha)	Food aid (000MT)	Food import-export ratio (value)	Volume of fert. (MT)	Volume of improved seed (MT)	Population in hunger (000)	Rainfall (mm)	Volume of food import (000MT)
Mean	5565	6504	1.17	791	28.88	136788	5981.83	5982	964	366910
Maximum	7731	9578	1.34	1249	62.13	291948	18536.34	18536	1383	846775
Minimum	4034	3836	0.88	202	1.58	8089	907.98	908	485	12828
Std Dev.	1256	1664	0.13	336	18.23	102591	5054.65	5055	249	236381
Skewness	1	1	-0.71	0	0.14	0	1.29	1	0	0
Kurtosis	2	2	2.65	2	2.08	1	3.56	4	2	2
CV	**23**	**26**	**10.79**	**43**	**63.10**	**75**	**84.50**	**85**	**26**	**64**
Jarque-Bera	3	2	2.14	1	0.69	2	6.73	7	0	1
Probability	0	0	0.34	1	0.71	0	0.03	0	1	1

Source: Own computations based on CSA and National Bank series of reports

Note: CV is coefficient of variation: CV = (Standard deviation ÷ arithmetic mean)*100. It measures the degree of variability of a variable: the larger the figure. the more variable the variable is. As may be seen in the tables, the CV of the prices of the cereals is closer to 50% and that of the quantities of improved seeds and chemical fertilizer is over 75%. These are very high compared to the experience even of sub-Saharan countries.

Table A6.2: Summary of Unit Root Tests

Descriptions of test	ADF (level)	ADF I(1)	ADF I(2)	PP (level)	PP I(1)
Tef price	-1.09	-3.06	-6.28*	-1.15	-3.77*
Wheat price	-2.14	-3.12	-5.12*	-3.14	-4.17*
Maize price	-2.39	-3.13	-5.34*	-1.59	-3.61*
Weighted cereal price (CP)	-3.00	-5.07*	---	-3.14	-5.4*
Relative product-input price ratio (RPR)	-2.67	-4.10*	---	-2.96	-5.07*
Volume of cereal output (CY)	-3.19	-5.02*	---	-2.45	-3.59*
Cereal area (CA)	-2.24	3.37*	---	-2.09	3.69*
Weighted volume of fertilizer & improved seed (MDI)	-1.91	-1.99	-3.69*	-2.06	-3.67*

Source: Own computation from CSA and own data
Notes: (*) indicates rejection of the hypothesis that unit root is present based on the Mackinnon critical values (Mackinnon. 1991); trends and intercept were included in the tests. and first differenced lag 1 in ADF and truncation lag 2 in PP were assumed.

Table A6.3: Pairwise Granger Causality Tests (1980-2003)

Null Hypothesis:	F-statistic	Probability
CY does not Granger-cause MDI	9.708	0.002**
CY does not Granger-cause Lag MDI	10.245	0.003**
CY does not Granger-cause RPR	1.004	0.391
Lag CY does not Granger-cause MDI	2.741	0.097*
Lag CY does not Granger-cause RPR	1.049	0.376
MDI does not Granger-cause CY	6.715	0.008**
MDI does not Granger-cause RPR	1.545	0.250
Lag MDI does not Granger-cause CY	8.734	0.003**
Lag MDI does not Granger-cause RPR	2.990	0.083*
RPR does not Granger-cause CY	6.300	0.011**
RPR does not Granger-cause MDI	3.071	0.081*
RPR does not Granger-cause Lag CY	7.724	0.005**
Lag RPR does not Granger-cause CY	6.300	0.011**
Lag RPR does not Granger-cause MDI	3.071	0.081*

Source: Own computation
Note: Two-level lags were considered.
* Significant at the 10% confidence level
** Significant at the 5% confidence level

Development Economics and Policy

Series edited by Franz Heidhues and Joachim von Braun

Band 1 Andrea Fadani: Agricultural Price Policy and Export and Food Production in Cameroon. A Farming Systems Analysis of Pricing Policies. The Case of Coffee-Based Farming Systems. 1999.

Band 2 Heike Michelsen: Auswirkungen der Währungsunion auf den Strukturanpassungsprozeß der Länder der afrikanischen Franc-Zone. 1995.

Band 3 Stephan Bea: Direktinvestitionen in Entwicklungsländern. Auswirkungen von Stabilisierungsmaßnahmen und Strukturreformen in Mexiko. 1995.

Band 4 Franz Heidhues / François Kamajou: Agricultural Policy Analysis – Proceedings of an International Seminar, held at the University of Dschang, Cameroon on May 26 and 27 1994, funded by the European Union under the Science and Technology Program (STD). 1996.

Band 5 Elke M. Förster: Protection or Liberalization? A Policy Analysis of the Korean Beef Sector. 1996.

Band 6 Gertrud Schrieder: The Role of Rural Finance for Food Security of the Poor in Cameroon. 1996.

Band 7 Nestor R. Ahoyo Adjovi: Economie des Systèmes de Production intégrant la Culture de Riz au Sud du Bénin: Potentialités, Contraintes et Perspectives. 1996.

Band 8 Jenny Müller: Income Distribution in the Agricultural Sector of Thailand. Empirical Analysis and Policy Options. 1996.

Band 9 Michael Brüntrup: Agricultural Price Policy and its Impact on Production, Income, Employment and the Adoption of Innovations. A Farming Systems Based Analysis of Cotton Policy in Northern Benin. 1997.

Band 10 Justin Bomda: Déterminants de l'Epargne et du Crédit, et leurs Implications pour le Développement du Système Financier Rural au Cameroun. 1998.

Band 11 John M. Msuya: Nutrition Improvement Projects in Tanzania: Implementation, Determinants of Performance, and Policy Implications. 1998.

Band 12 Andreas Neef: Auswirkungen von Bodenrechtswandel auf Ressourcennutzung und wirtschaftliches Verhalten von Kleinbauern in Niger und Benin. 1999.

Band 13 Susanna Wolf (ed.): The Future of EU-ACP Relations. 1999.

Band 14 Franz Heidhues / Gertrud Schrieder (eds.): Romania – Rural Finance in Transition Economies. 2000.

Band 15 Katinka Weinberger: Women's Participation. An Economic Analysis in Rural Chad and Pakistan. 2000.

Band 16 Christof Batzlen: Migration and Economic Development. Remittances and Investments in South Asia: A Case Study of Pakistan. 2000.

Band 17 Matin Qaim: Potential Impacts of Crop Biotechnology in Developing Countries. 2000.

Band 18 Jean Senahoun: Programmes d'ajustement structurel, sécurité alimentaire et durabilité agricole. Une approche d'analyse intégrée, appliquée au Bénin. 2001.

Band 19 Torsten Feldbrügge: Economics of Emergency Relief Management in Developing Countries. With Case Studies on Food Relief in Angola and Mozambique. 2001.

Band 20 Claudia Ringler: Optimal Allocation and Use of Water Resources in the Mekong River Basin: Multi-Country and Intersectoral Analyses. 2001.

Band 21 Arnim Kuhn: Handelskosten und regionale (Des-)Integration. Russlands Agrarmärkte in der Transformation. 2001.

Band 22 Ortrun Anne Gronski: Stock Markets and Economic Growth. Evidence from South Africa. 2001.

Band 23 Patrick Webb / Katinka Weinberger (eds.): Women Farmers. Enhancing Rights, Recognition and Productivity. 2001.

Band 24 Mingzhi Sheng: Lebensmittelkonsum und -konsumtrends in China. Eine empirische Analyse auf der Basis ökonometrischer Nachfragemodelle. 2002.

Band 25 Maria Iskandarani: Economics of Household Water Security in Jordan. 2002.

Band 26 Romeo Bertolini: Telecommunication Services in Sub-Saharan Africa. An Analysis of Access and Use in the Southern Volta Region in Ghana. 2002.

Band 27 Dietrich Müller-Falcke: Use and Impact of Information and Communication Technologies in Developing Countries' Small Businesses. Evidence from Indian Small Scale Industry. 2002.

Band 28 Wolfram Erhardt: Financial Markets for Small Enterprises in Urban and Rural Northern Thailand. Empirical Analysis on the Demand for and Supply of Financial Services, with Particular Emphasis on the Determinants of Credit Access and Borrower Transaction Costs. 2002.

Band 29 Wensheng Wang: The Impact of Information and Communication Technologies on Farm Households in China. 2002.

Band 30 Shyamal K. Chowdhury: Institutional and Welfare Aspects of the Provision and Use of Information and Communication Technologies in the Rural Areas of Bangladesh and Peru. 2002.

Band 31 Annette Luibrand: Transition in Vietnam. Impact of the Rural Reform Process on an Ethnic Minority. 2002.

Band 32 Felix Ankomah Asante: Economic Analysis of Decentralisation in Rural Ghana. 2003.

Band 33 Chodechai Suwanaporn: Determinants of Bank Lending in Thailand: An Empirical Examination for the Years 1992 to 1996. 2003.

Band 34 Abay Asfaw: Costs of Illness, Demand for Medical Care, and the Prospect of Community Health Insurance Schemes in the Rural Areas of Ethiopia. 2003.

Band 35 Gi-Soon Song: The Impact of Information and Communication Technologies (ICTs) on Rural Households. A Holistic Approach Applied to the Case of Lao People's Democratic Republic. 2003.

Band 36 Daniela Lohlein: An Economic Analysis of Public Good Provision in Rural Russia. The Case of Education and Health Care. 2003.

Band 37 Johannes Woelcke. Bio-Economics of Sustainable Land Management in Uganda. 2003.

Band 38 Susanne M. Ziemek: The Economics of Volunteer Labor Supply. An Application to Countries of a Different Development Level. 2003.

Band 39 Doris Wiesmann: An International Nutrition Index. Concept and Analyses of Food Insecurity and Undernutrition at Country Levels. 2004.

Band 40 Isaac Osei-Akoto: The Economics of Rural Health Insurance. The Effects of Formal and Informal Risk-Sharing Schemes in Ghana. 2004.

Band 41 Yuansheng Jiang: Health Insurance Demand and Health Risk Management in Rural China. 2004.

Band 42 Roukayatou Zimmermann: Biotechnology and Value-added Traits in Food Crops: Relevance for Developing Countries and Economic Analyses. 2004.

Band 43 F. Markus Kaiser: Incentives in Community-based Health Insurance Schemes. 2004.

Band 44 Thomas Herzfeld: *Corruption begets Corruption.* Zur Dynamik und Persistenz der Korruption. 2004.

Band 45 Edilegnaw Wale Zegeye: The Economics of On-Farm Conservation of Crop Diversity in Ethiopia: Incentives, Attribute Preferences and Opportunity Costs of Maintaining Local Varieties of Crops. 2004.

Band 46 Adama Konseiga: Regional Integration Beyond the Traditional Trade Benefits: Labor Mobility contribution. The Case of Burkina Faso and Côte d'Ivoire. 2005.

Band 47 Beyene Tadesse Ferenji: The Impact of Policy Reform and Institutional Transformation on Agricultural Performance. An Economic Study of Ethiopian Agriculture. 2005.

www.peterlang.de